SpringerBriefs in Applied Sciences
and Technology

Manufacturing and Surface Engineering

Series Editor

J. Paulo Davim

For further volumes:
http://www.springer.com/series/10623

C. V. Nielsen · W. Zhang · L. M. Alves
N. Bay · P. A. F. Martins

Modeling of Thermo-Electro-Mechanical Manufacturing Processes

Applications in Metal Forming and Resistance Welding

 Springer

C. V. Nielsen
Department of Mechanical Engineering
Technical University of Denmark
Kongens Lyngby
Denmark

N. Bay
Department of Mechanical Engineering
Technical University of Denmark
Kongens Lyngby
Denmark

W. Zhang
SWANTEC Software and Engineering ApS
Kongens Lyngby
Denmark

P. A. F. Martins
IDMEC, Instituto Superior Técnico
Technical University of Lisbon
Lisbon
Portugal

L. M. Alves
IDMEC, Instituto Superior Técnico
Technical University of Lisbon
Lisbon
Portugal

ISBN 978-1-4471-4642-1   ISBN 978-1-4471-4643-8   (eBook)
DOI 10.1007/978-1-4471-4643-8
Springer London Heidelberg New York Dordrecht

Library of Congress Control Number: 2012948188

Printed on acid-free paper

Springer is part of Springer Science+Business Media (www.springer.com)

Contents

Chapter 1
Introduction

Taking a general view of the present state of the art in terms of modeling and computation of manufacturing processes it appears that the finite element flow formulation is one of the most widespread numerical methodologies for the analysis of complex, industrial metal forming and resistance welding processes.

The finite element flow formulation is capable of providing very efficient computer programs that can take into account the practical non-linearities in the geometry and material properties as well as the contact change typical of the interaction between workpieces and tools to produce accurate predictions of displacements, strain rates, strains, stresses, damage, temperature and current density, among other variables.

Nowadays, commercial computer programs based on the finite element flow formulation such as DEFORM, FORGE, QFORM and eesy-2-form are standard engineering tools for designing and optimizing metal forming processes. SORPAS, also based on the flow formulation, is the reference commercial computer program for industrial resistance welding processes.

In contrast to the active role performed by manufacturing research groups during the theoretical and numerical developments of finite element computer programs that were produced during the 1980s and 1990s [1], current practice seems to indicate a total or near-total engagement of the majority of these groups on applications rather than on developments. A critical gap was formed between the developers of the computer programs and the users having the know-how on the metal forming and resistance welding technologies.

This book deals with the above-mentioned gap between developers and users and it is designed with a three-fold objective:

- to provide readers with a better understanding of the fundamental ingredients in plasticity, heat transfer and electricity that are necessary to develop and proper utilize computer programs based on the finite element flow formulation;
- to discuss computer implementation of a wide range of theoretical and numerical subjects related to mesh generation, contact algorithms, elasticity,

C. V. Nielsen et al., *Modeling of Thermo-Electro-Mechanical Manufacturing Processes*, SpringerBriefs in Applied Sciences and Technology, DOI: 10.1007/978-1-4471-4643-8_1, © The Author(s) 2013

anisotropic constitutive equations, solution procedures and parallelization of equation solvers, among others;
- to draw from the fundamentals of the flow formulation to aspects of accuracy, reliability and validation of numerical modeling by presenting industrial examples related to the development of new products and to the optimization and increasing know-how of existing products and processes.

Besides special purpose examples taken from metal forming applications, which are included for enriching the presentation of specific theoretical and numerical contents, the last chapter is focused on industrial applications of joining technologies by metal forming and resistance welding. Joining technologies combine main aspects of plasticity, electricity and heat transfer for assembling individual components to complete and useful end products and offer the possibility of selecting industrial applications that deal with state-of-the-art engineering concepts that, although not being commonly available in the open research literature, are a good option for establishing communication links between the developers of finite element computer programs and the professional engineers having the know-how on metal forming and resistance welding technologies.

In order to get full benefit of the industrial applications included in the book, those readers who are not familiar with metal forming or resistance welding technologies may enjoy reading and exploring the reference handbooks provided by The American Society of Metals [2] and The Resistance Welding Manufacturers' Association [3].

References

1. Kobayashi S, Oh SI, Altan T (1989) Metal forming and the finite element method. Oxford University Press, Oxford
2. American Society of Metals (1988) Forming and forging. ASM International, Materials Park
3. Resistance Welding Manufacturers' Association (2003) Resistance welding manual. RWMA, Philadelphia

Chapter 2
Finite Element Formulations

The governing equations for problems solved by the finite element method are typically formulated by partial differential equations in their original form. These are rewritten into a weak form, such that domain integration can be utilized to satisfy the governing equations in an average sense. A functional Π is set up for the system, typically describing the energy or energy rate and implying that the solution can be found by minimization. For a generic functional, this is written as

$$\frac{\partial \Pi}{\partial u} = \frac{\partial}{\partial u}\left(\int_V f(x_i, u_i)\, dV\right) = \frac{\partial}{\partial u}\left(\sum_j f(x_i, u_i)\, \Delta V_j\right) = 0 \qquad (2.1)$$

where the functional is a function of the coordinates x_i and the primary variable u_i being e.g. displacements or velocities for mechanical problems depending on the formulation. The domain integration is approximated by a summation over a finite number of elements discretizing the domain. Figure 2.1 illustrates a three-dimensional domain discretized by hexahedral elements with eight nodes. The variables are defined and solved in the nodal points, and evaluation of variables in the domain is performed by interpolation in each element. Shared nodes give rise to an assembly of elements into a global system of equations of the form

$$\mathbf{Ku} = \mathbf{f} \qquad (2.2)$$

where $\mathbf{K}$ is the stiffness matrix, $\mathbf{u}$ is the primary variable and $\mathbf{f}$ is the applied load, e.g. stemming from applied tractions F on a surface S_F in Fig. 2.1. The system of equations (2.2) is furthermore subject to essential boundary conditions, e.g. prescribed displacements or velocities u along a surface S_U.

The basic aspects of available finite element formulations in terms of modeling and computation are briefly reviewed in this chapter. This will support the choice of formulation to be detailed and applied in the remaining chapters, where an electro-thermo-mechanical finite element formulation is presented

C. V. Nielsen et al., *Modeling of Thermo-Electro-Mechanical Manufacturing Processes*, SpringerBriefs in Applied Sciences and Technology, DOI: 10.1007/978-1-4471-4643-8_2, © The Author(s) 2013

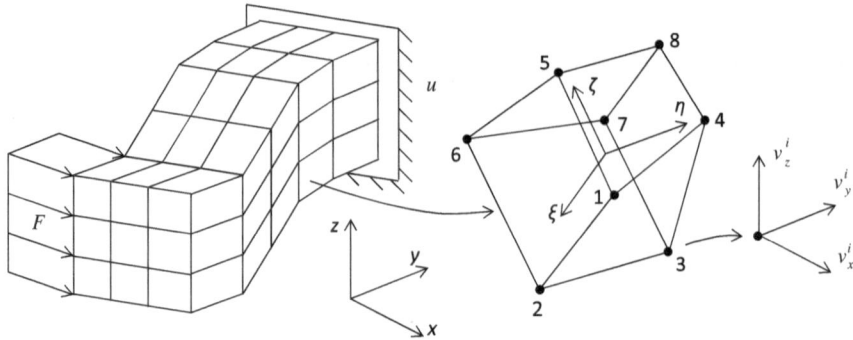

Fig. 2.1 Illustration of three-dimensional finite element model composed of isoparametric, hexahedral elements with eight nodes. Each node has three degrees of freedom for representation of vector fields and one degree of freedom for representation of scalar fields

together with a range of aspects to complete a computer program capable of modeling manufacturing processes such as metal forming and resistance welding. This chapter is focused on the mechanical formulations because they represent major differences and because the mechanical model plays a central role in the overall modeling strategy. From a process point of view the mechanical model is responsible for material flow, contact and stress distribution, and from a computational point of view is responsible for the largest amount of CPU time. In addition, the overall structure of the presented computer program is built upon the mechanical formulation with the remaining thermal and electrical modules integrated.

One fundamental difference between the finite element formulations is the governing equilibrium equation, being either quasi-static or dynamic in the modeling of manufacturing processes. Another fundamental choice to cover is the material model suited for describing the materials under consideration, bearing in mind the process to simulate and thereby the expected range of deformation and deformation rate. The available constitutive models to utilize in the material description are rigid-plastic/viscoplastic and elasto-plastic/viscoplastic.

Table 2.1, after Tekkaya and Martins [1], provides an overview of the quasi-static formulations and the dynamic formulation. The quasi-static formulations are represented by the flow formulation and the solid formulation, distinguishable by the underlying constitutive equations. The following two sections are devoted to give a brief overview of the quasi-static and dynamic formulations including their advantages and disadvantages.

Presentation of the quasi-static and dynamic formulations follows the general outline given by Tekkaya and Martins [1] and additional information can be found in major reference books by Zienkiewicz and Taylor [2], Banabic et al. [3], Wagoner and Chenot [4] and Dunne and Petrinic [5].

Table 2.1 Overview of finite element formulations and commercial computer programs applied in the metal forming industry

	Quasi-static formulations		Dynamic formulation
	Flow formulation	Solid formulation	
Equilibrium equation:	Quasi-static	Quasi-static	Dynamic
Constitutive equations:	Rigid-plastic/ viscoplastic	Elasto-plastic/ viscoplastic	Elasto-plastic/ viscoplastic
Main structure:	Stiffness matrix and force vector	Stiffness matrix and force vector	Mass and damping matrices and internal and external force vectors
Solution scheme[a]:	Implicit	Implicit	Explicit
Size of incremental step:	Large	Medium to large	Very small
CPU time per incremental step:	Medium	Medium to long	Very short
Time integration scheme[b]:	Explicit	Implicit	Explicit
Accuracy of the results (stress and strain distributions):	Medium to high	High	Medium to low
Springback and residual stresses:	No (although the basic formulation can be modified to include elastic recovery)	Yes	Yes/no
Commercial FEM computer programs related to metal forming	FORGE[c], DEFORM[c], QFORM, eesy-2-form	Abaqus (implicit), Simufact.forming, AutoForm, Marc	Abaqus (explicit), DYNA3D, PAM-STAMP

[a]Explicit/implicit if the residual force is not/is minimized at each incremental step.
[b]Explicit/implicit if the algorithm does not/does need the values of the next time step to compute the solution.
[c]Elasto-plastic options available.

2.1 Quasi-Static Formulations

The quasi-static formulations are governed by the static equilibrium equation, which in the absence of body forces takes the following form,

$$\sigma_{ij,j} = 0 \qquad (2.1.1)$$

where $\sigma_{ij,j}$ denotes the partial derivatives of the Cauchy stress tensor with respect to the Cartesian coordinates x_j. This equation expresses the equilibrium in the current configuration, i.e. in the mesh following the deformation.

By employing the Galerkin method, it is possible to write an integral form of Eq. (2.1.1) that fulfills the equilibrium in an average sense over the entire domain instead of satisfying the equilibrium point-wise. This formulation allows domain integration to substitute the more tedious solution of the original differential equations. The integral over domain volume V is

$$\int_V \sigma_{ij,j} \delta u_i dV = 0 \qquad (2.1.2)$$

with δu_i being an arbitrary variation in the primary unknown u_i, which is either displacement or velocity depending on the implementation. Displacement is the primary unknown in rate independent formulations and velocity is the primary unknown in rate dependent formulations.

Applying integration by parts in Eq. (2.1.2), followed by the divergence theorem and taking into account the natural and essential boundary conditions, it is possible to rewrite Eq. (2.1.2) as follows,

$$\int_V \sigma_{ij} (\delta u_i)_{,j} dV - \int_S t_i \delta u_i dS = 0 \qquad (2.1.3)$$

where $t_i = \sigma_{ij} n_j$ denotes the tractions with direction of the unit normal vector n_j applied on the boundary surface S. Equation (2.1.3) is the "weak variational form" of Eq. (2.1.1) because the static governing equilibrium equations are now only satisfied under weaker continuity requirements.

The above listed equations together with appropriate constitutive equations enable quasi-static finite element formulations to be defined by means of the following matrix set of non-linear equations,

$$\mathbf{K}^t \mathbf{u}^t = \mathbf{F}^t \qquad (2.1.4)$$

which express the equilibrium condition at the instant of time t through the stiffness matrix $\mathbf{K}$, the generalized force vector $\mathbf{F}$ resulting from the loads, pressure and friction stresses applied on the boundary. The equation system is non-linear due to the stiffness matrix's dependency of the primary unknown $\mathbf{u}$ to geometry and material properties.

The quasi-static finite element formulations utilized in the analysis of metal forming and resistance welding processes are commonly implemented in conjunction with implicit solution schemes. The main advantage of implicit schemes over alternative solutions based on explicit procedures is that equilibrium is checked at each increment of time by means of iterative procedures to minimize the residual force vector $\mathbf{R(u)}$, which is computed as follows in iteration number n,

$$\mathbf{R}_n^t = \mathbf{K}_{n-1}^t \mathbf{u}_n^t - \mathbf{F}^t \qquad (2.1.5)$$

The non-linear set of equations (2.1.4), derived from the quasi-static implicit formulations, can be solved by different numerical techniques such as the direct iteration (also known as "successive replacement") and the Newton–Raphson methods. In the direct iteration method, the stiffness matrix is evaluated for the displacements of the previous iteration in order to reduce Eq. (2.1.4) to a linear set of equations. The method is iterative and converges linearly and unconditionally towards the solution during the earlier stages of the iteration procedure but becomes slow as the solution is approached. The standard Newton-Raphson method is an alternative iterative method based on a linear expansion of the residual $\mathbf{R}(\mathbf{u})$ near the velocity estimate at the previous iteration,

$$\mathbf{R}_n^t = \mathbf{R}_{n-1}^t + \left[\frac{\partial \mathbf{R}}{\partial \mathbf{u}}\right]_{n-1}^t \Delta \mathbf{u}_n^t = 0 \qquad (2.1.6a)$$

$$\mathbf{u}_n^t = \mathbf{u}_{n-1}^t + \Delta \mathbf{u}_n^t \qquad (2.1.6b)$$

This procedure is only conditionally convergent, but converges quadratically in the vicinity of the exact solution. The iterative procedures are designed in order to minimize the residual force vector $\mathbf{R}(\mathbf{u})$ to within a specified tolerance. Control and assessment is performed by means of appropriate convergence criteria.

The main advantage of the quasi-static implicit finite element formulations is that equilibrium conditions are checked at each increment of time in order to minimize the residual force vector $\mathbf{R}(\mathbf{u})$ to within a specified tolerance.

The main drawbacks in the quasi-static implicit finite element formulations are summarized as follows:

- Solution of linear systems of equations is required during each iteration;
- High computation times and high memory requirements;
- Computation time depends quadratically on the number of degrees of freedom if a direct solver is utilized, and with the Newton-Raphson method the solution is only conditionally convergent;
- The stiffness matrix is often ill-conditioned, which can turn the solution procedure unstable and deteriorate the performance of iterative solvers;
- Difficulties in dealing with complex non-linear contact and tribological boundary conditions are experienced, and that often leads to convergence problems.

2.2 Dynamic Formulation

The dynamic finite element formulation is based on the dynamic equilibrium equation in the current configuration, here written in the absence of body forces with the inertia term expressed through the mass density ρ and the acceleration $\ddot{u}_i$,

$$\sigma_{ij,j} - \rho \ddot{u}_i = 0 \qquad (2.2.1)$$

Applying a mathematical procedure similar to that described in the previous section results in the following weak variational form of Eq. (2.2.1),

$$\int_V \rho \ddot{u}_i \delta u_i dV + \int_V \sigma_{ij} (\delta u_i)_{,j} dV - \int_S t_i \delta u_i dS = 0 \qquad (2.2.2)$$

The above equation enables dynamic finite element formulations to be represented by the following matrix set of non-linear equations,

$$\mathbf{M}^t \ddot{\mathbf{u}}^t + \mathbf{F}_{\text{int}}^t = \mathbf{F}^t \qquad (2.2.3)$$

which express the dynamic equilibrium condition at the instant of time t. The symbol $\mathbf{M}$ denotes the mass matrix, $\mathbf{F}_{int} = \mathbf{K}\mathbf{u}$ is the vector of internal forces resulting from the stiffness, and $\mathbf{F}$ is the generalized force vector.

The non-linear set of equations (2.2.3), derived from the dynamic formulation, is commonly solved by means of an explicit central difference time integration scheme,

$$\mathbf{M}^t \left(\frac{\dot{\mathbf{u}}^{t+1/2} - \dot{\mathbf{u}}^{t-1/2}}{\Delta t^{t+1/2}} \right) + \mathbf{F}_{int}^t = \mathbf{F}^t \tag{2.2.4a}$$

$$\dot{\mathbf{u}}^{t+1/2} = \left(\mathbf{M}^t\right)^{-1} \left(\mathbf{F}^t - \mathbf{F}_{int}^t\right) \Delta t^{t+1/2} + \dot{\mathbf{u}}^{t-1/2} \tag{2.2.4b}$$

$$\mathbf{u}^{t+1} = \mathbf{u}^t + \dot{\mathbf{u}}^{t+1/2}\Delta t^{t+1} \tag{2.2.4c}$$

If the mass matrix $\mathbf{M}$ in Eq. (2.2.4a, b) is diagonalized (or lumped) its inversion is trivial, and the system of differential equations decouples. Its overall solution can then be performed independently and very fast for each degree of freedom. Further reductions of the computation time per increment of time stem from utilization of reduced integration schemes that are often applied even to the deviatoric parts of the stiffness matrix, and finally numerical actions related to mass scaling and load factoring contribute. Load factoring is described ahead.

Additional computational advantages result from the fact that dynamic explicit schemes, unlike quasi-static implicit schemes, do not check equilibrium requirements at the end of each increment of time. The analogy between the dynamic equilibrium equation (2.2.1) and the ideal mass-spring vibrating system allows concluding that explicit central difference time integration schemes (frequently referred as explicit integration schemes) are conditionally stable whenever the size of the increment of time Δt satisfies

$$\Delta t \leq \frac{L_e}{\sqrt{E/\rho}} = \frac{L_e}{c_e} \tag{2.2.5}$$

where L_e is the typical size of the finite elements discretizing the domain, E is the elasticity modulus and c_e is the velocity of propagation of a longitudinal wave in the material. In case of metal forming applications, the stability condition Eq. (2.2.5) requires the utilization of very small increments of time Δt, say microseconds, and millions of increments to finish a simulation because industrial metal forming processes usually take several seconds to be accomplished. This is the reason why computer programs often make use of the following numerical actions in order to increase the increment of time Δt and, consequently, reducing the overall computation time:

- Diagonalization of the mass matrix;
- Mass scaling—by increasing the density of the material and thus artificially reducing the speed c_e of the longitudinal wave;
- Load factoring—by changing the rate of loading through an artificial increase in the velocity of the tooling as compared to the real forming velocity;
- Reduced integration of the deviatoric part of the stiffness matrix, which is usually fully integrated.

The above-mentioned numerical actions can artificially add undesirable inertia effects, and it is therefore necessary to include a damping term $\mathbf{C}^t\dot{\mathbf{u}}^t$ in (2.2.3),

$$\mathbf{M}^t\ddot{\mathbf{u}}^t + \mathbf{C}^t\dot{\mathbf{u}}^t + \mathbf{F}_{\text{int}}^t = \mathbf{F}^t \qquad (2.2.6)$$

The damping term $\mathbf{C}^t\dot{\mathbf{u}}^t$ is not only necessary because of the above-mentioned numerical actions to reduce the computation time but also to ensure fast convergence of the solution towards the static solution describing the actual process.

This turns dynamic explicit formulations into close resemblance with damped mass-spring vibrating systems and justifies the reason why these formulations loose efficiency whenever the material is strain-rate sensitive or thermo-mechanical phenomena need to be taken into consideration.

The main advantages of the dynamic explicit formulations are:

- Computer programs are robust and do not present convergence problems;
- The computation time depends linearly on the number of degrees of freedom while in alternative quasi-static implicit schemes the dependency is more than linear (in case of iterative solvers) and up to quadratic (in case of direct solvers).

The main drawbacks of the dynamic explicit formulation can be summarized as follows:

- Utilization of very small time increments;
- Equilibrium after each increment of time is not checked;
- Assignment of the system damping is rather arbitrary;
- The formulation needs experienced users for adequately designing the mesh and choosing the scaling parameters for mass, velocity and damping. Otherwise it may lead to inaccurate solutions for the deformation, prediction of forming defects and distribution of the major field variables within the workpiece;
- Springback calculations are very time consuming and may lead to errors. This specific problem is frequently overtaken by combining dynamic explicit with quasi-static implicit analysis.

The last two drawbacks apply if the dynamic explicit formulations are used in the "high-speed-mode".

References

1. Tekkaya AE, Martins PAF (2009) Accuracy, reliability and validity of finite element analysis in metal forming: a user's perspective. Eng Comput 26:1026–1055
2. Zienkiewicz OC, Taylor RL (2000) The finite element method. Butterworth-Heinemann, Oxford
3. Banabic D, Bunge H-J, Pöhlandt K, Tekkaya AE (2000) Formability of metallic materials. Springer, Berlin
4. Wagoner RH, Chenot JL (2001) Metal forming analysis. Cambridge University Press, Cambridge
5. Dunne F, Petrinic N (2006) Introduction to computational plasticity. Oxford University Press, Oxford

Chapter 3
Coupled Finite Element Flow Formulation

This chapter presents a coupled finite element approach for thermo-mechanical modeling of metal forming and for electro-thermo-mechanical modeling of resistance welding. The finite element approach is based on the flow formulation which was described in Chap. 2 as one of the implicit quasi-static formulations.

Direct comparison of the performance achieved with the implicit quasi-static formulations based on flow and solid approaches (refer to Table 2.1) are provided by Boer et al. [1] and Kobayashi et al. [2], who emphasize the advantages of the flow approach in modeling the mechanical response (plastic flow) of materials undergoing large deformations.

3.1 State-of-the-Art

Taking a general view to the bibliographic retrieval by Brännberg and Mackerle [3] and Mackerle [4, 5] it appears that the finite element flow formulation is one of the most widespread numerical methodologies for the analysis of metal forming processes.

In the flow formulation, the material is treated in a similar way to an incompressible fluid. Rigid-plastic/viscoplastic constitutive laws are utilized and the elastic response is neglected, simplifying the problem and offering additional computational advantages. The computer programs based on the flow formulation can successfully take into account the non-linearities in the geometry and material properties as well as the contact changes typical of metal forming and resistance welding processes to produce accurate predictions of plastic flow, temperature, current density and microstructure.

In order to calculate temperatures and its resulting effects, the flow formulation is coupled with heat transfer analysis to achieve complete thermo-mechanical modeling. In resistance welding this coupling is further extended to include electrical analysis with special treatment of contact interfaces and to account for Joule

C. V. Nielsen et al., *Modeling of Thermo-Electro-Mechanical Manufacturing Processes*, SpringerBriefs in Applied Sciences and Technology, DOI: 10.1007/978-1-4471-4643-8_3, © The Author(s) 2013

heating. The extended model is electro-thermo-mechanically coupled and enables utilization in a wide range of manufacturing applications by industry, research and education institutions with the aim of:

- Developing new products and processes in shorter time;
- Optimizing existing products and processes by cost and quality;
- Increasing process understanding and strengthening technological know-how;
- Performing more efficient experimentation by providing starting parameters and support to the analyses.

Modeling and simulation of manufacturing processes are tools to better understand and thereby solve new problems arising when forming or joining new materials and geometries. In what concerns resistance welding, Singh [6] pointed out that simulation cannot replace or substitute ingenuity or creativeness, but it can help in gaining understanding of the process, and thus it can reduce the amount of time spent during development.

The interaction with industry has been the motivation for applying and continuously developing the finite element flow formulation for manufacturing applications over the past decades. A brief overview of the previous research in the field is given in what follows with the aim of providing a timeline of the major contributions and identifying the current state-of-the-art.

The finite element flow formulation was originally developed by Lee and Kobayashi [7], Cornfield and Johnson [8] and Zienkiewicz and Godbole [9] during the 1970s with the aim of simulating metal forming processes. During the 1980s, the flow formulation was primarily set up for modeling two-dimensional bulk forming processes and such efforts gave rise to the development of a first generation of commercial software with applicability limited to plane strain and axisymmetric conditions. Even so, authors such as Altan and Knoerr [10] were able to report case studies in which the two-dimensional constraint was ingeniously stretched out in order to obtain useful information regarding three-dimensional metal forming applications.

In order to extend applicability of the flow formulation to modeling conditions involving more than the mechanical behavior alone, a thermal model was introduced to simulate thermo-mechanical manufacturing processes. The first attempt to handle a coupled thermo-mechanical metal forming process was made by Zienkiewicz et al. [11] who used a finite element iterative procedure to solve the material flow for a given distribution of temperature, in conjunction with the heat transfer phenomenon, during plane strain extrusion. Later, Zienkiewicz et al. [12, 13] modified the procedure to allow the temperature distribution within the workpiece to be obtained simultaneously with the solution of the velocity field. The modification, commonly known as "direct coupled thermo-mechanical" was applied to solve steady-state extrusion and rolling. The heat exchange with the tools was either neglected, as in the case of the extrusion problem, or taken into account by imposing a constant temperature on the tools, as in the case of steady-state rolling.

Direct coupled thermo-mechanical finite element algorithms were further developed by Rebelo and Kobayashi [14, 15] to allow the numerical simulation of non-steady-state metal forming processes. The technique was applied to solid cylinder and ring compression testing.

As regards resistance welding, early contributions, being analytical or numerical, were focused on the temperature field under a given voltage potential. Nied [16] was the first to present electro-thermo-mechanical modeling of spot welding by finite elements using the commercial program ANSYS. The study was performed in two dimensions with assumed Hertzian elastic contact. Contact conditions are crucial for the numerical simulation of resistance welding due to dynamically developing contact area and Nied [16] addressed this problem by means of surface elements that were capable of supporting compressive stresses, but not tensile stresses. Relative sliding was allowed assuming frictionless contact and electrical and thermal properties were included in the aforementioned surface elements.

The work of Nied [16] was the first numerical simulation of resistance welding being so complete. Subsequent published work in the field was also based on commercial finite element computer programs for general purpose modeling, e.g. Zhu et al. [17] modeled projection welding of an automotive door hinge with two projections to a sheet metal by means of a two-dimensional analysis based on ANSYS.

Newer developments in computers and reduction in the associated computational costs are presently extending the availability and effectiveness of finite element software to simulate three-dimensional manufacturing processes. As a consequence, complex processes are now being simulated precisely without the need to take advantage of possible geometrical and material flow simplifications. A detailed survey of the state-of-the-art regarding numerical simulation of metal forming processes is given by Brännberg and Mackerle [3] and Mackerle [4, 5].

In resistance welding state-of-the-art is the prediction of weld parameters in spot welding with the only input being the geometries of sheets and electrodes as well as the desired weld nugget size (refer, for example, to the presentation of weld planning in SORPAS by Zhang [18]). SORPAS is a commercial finite element program dedicated to simulation and optimization of resistance projection and spot welding. The program is based on the finite element flow formulation and has been primarily developed for axisymmetric and plane-strain industrial applications. However, recent developments by Nielsen et al. [19] have extended the capabilities of SORPAS to simulate complex three-dimensional resistance welding applications.

Most of the scientific and numerical ingredients that are necessary to develop computer programs (discretization procedures, solution techniques, contact algorithms and remeshing schemes) are drawing continuous attention in the literature with the purpose of enhancing the overall standards of accuracy, performance and robustness of existing computer software. Despite the many contributions in literature, it is becoming more evident that industrial, and to some degree also academic, use of finite element modeling is relying on existing commercial computer programs, where the user naturally has limited access to implementation details and the physical models behind the calculations. This leads to potential pitfalls if the users are not aware of the best use and potential limitations of such programs, as recently discussed by Tekkaya and Martins [20].

This chapter is aimed to describe the fundamentals and numerical implementation of the thermo-mechanical and electro-thermo-mechanical coupled approaches that are available in academic (e.g. I-Form [21]) and commercial (e.g. SORPAS [18]) computer programs that are based on the finite element flow formulation.

3.2 Theoretical Background

3.2.1 Plastic Flow

The flow formulation is based on the quasi-static equilibrium equations, which in the absence of body forces and after some mathematical treatment that takes into consideration the natural and essential boundary conditions, can be written as (refer to Eq. (2.1.3) in Chap. 2).

$$\int_V \sigma_{ij}\delta\dot{\varepsilon}_{ij}dV - \int_S t_i\,\delta u_i dS = 0 \qquad (3.2.1.1)$$

where V is the domain volume, S is the boundary surface where tractions $t_i = \sigma_{ij}n_j$ are applied and $\dot{\varepsilon}_{ij}$ are the components of the strain rate tensor,

$$\dot{\varepsilon}_{ij} = \tfrac{1}{2}\left(u_{i,j} + u_{j,i}\right) \qquad (3.2.1.2)$$

In the flow formulation, velocities u_i are the primary unknown instead of displacements, there is no strain tensor and the stress σ_{ij} is directly related to the strain rate by means of rigid-plastic/viscoplastic constitutive equations.

In case of using the von Mises yield criterion, also called the "distortion energy criterion",

$$f\left(\sigma_{ij}\right) = \tfrac{1}{2}\sigma'_{ij}\sigma'_{ij} \qquad (3.2.1.3)$$

where f is the yield function and σ'_{ij} is the deviatoric stress tensor, the constitutive equations (also known as the "Levy–Mises equations") are written as

$$\dot{\varepsilon}_{ij} = \sigma'_{ij}\dot{\lambda} \qquad (3.2.1.4)$$

The proportionality factor $\dot{\lambda}$ in the above equation is given by

$$\dot{\lambda} = \frac{3}{2}\frac{\dot{\bar{\varepsilon}}}{\bar{\sigma}} \qquad (3.2.1.5)$$

with effective strain rate $\bar{\dot{\varepsilon}}$ and effective stress $\bar{\sigma}$ obtained from

$$\bar{\dot{\varepsilon}} = \sqrt{\tfrac{2}{3}} \left\{ \dot{\varepsilon}_{ij} \dot{\varepsilon}_{ij} \right\}^{\frac{1}{2}} \tag{3.2.1.6}$$

$$\bar{\sigma} = \sqrt{\tfrac{3}{2}} \left\{ \sigma'_{ij} \sigma'_{ij} \right\}^{\frac{1}{2}} \tag{3.2.1.7}$$

The variational principle associated with (3.2.1.1) requires that among admissible velocities u_i, satisfying the conditions of compatibility and incompressibility as well as the velocity boundary conditions, the actual solution gives the following functional a stationary value (minimum of the total energy rate),

$$\Pi = \int_V \bar{\sigma} \bar{\dot{\varepsilon}} \, dV - \int_S t_i u_i \, dS \tag{3.2.1.8}$$

where $\bar{\sigma} \bar{\dot{\varepsilon}} = \sigma_{ij} \dot{\varepsilon}_{ij}$ according to (3.2.1.6) and (3.2.1.7). Equation (3.2.1.1) corresponds to a zero first order variation of the total energy rate of the system (3.2.1.8) and is accordingly rewritten as follows,

$$\delta \Pi = \int_V \bar{\sigma} \, \delta \bar{\dot{\varepsilon}} \, dV - \int_S t_i \, \delta u_i \, dS = 0 \tag{3.2.1.9}$$

This is a weak form of the quasi-static equilibrium condition (2.1.1) because it lowers the continuity requirements on the stress field and allows solving the equilibrium condition by domain integration instead of the more tedious direct solving of differential equations.

In order to guarantee that the flow formulation is capable of providing a geometrically self-consistent velocity field that ensures the incompressibility condition it is necessary to ensure a zero first order variation of the functional Π (3.2.1.8), subject to a general constraint, $\dot{\varepsilon}_{kk} = 0$ over the entire domain. This can be done in several different ways, where the two most widespread techniques are based on the utilization of Lagrange multipliers (treating incompressibility as a mixed velocity–pressure approach) or penalties.

The utilization of a Lagrange multiplier λ_L, corresponding to the mean stress σ_m, modifies (3.2.1.9) to the following form,

$$\delta \Pi = \int_V \bar{\sigma} \, \delta \bar{\dot{\varepsilon}} \, dV + \int_V \lambda_L \, \delta \dot{\varepsilon}_{jj} \, dV + \int_V \delta \lambda_L \, \dot{\varepsilon}_{jj} \, dV - \int_S t_i \, \delta u_i \, dS = 0 \tag{3.2.1.10}$$

whereas the utilization of a penalty K, which is a large positive number related to the mean stress through $K \dot{\varepsilon}_{kk} = 2\sigma_m$, modifies (3.2.1.9) to the following form,

$$\delta \Pi = \int_V \bar{\sigma} \, \delta \bar{\dot{\varepsilon}} \, dV + K \int_V \dot{\varepsilon}_{ii} \, \delta \dot{\varepsilon}_{jj} \, dV - \int_S t_i \, \delta u_i \, dS = 0 \tag{3.2.1.11}$$

The advantage of the Lagrange multipliers is the exact solution, but on the expense of prolonged computation time due to additional unknowns in form of the mean stress $\sigma_m = \sigma_{kk}/3$ pressure terms.

The penalty approach does not introduce additional unknowns but suffers from a dilemma in the selection of the value of the penalty factor K. It has to be as large as possible to enforce incompressibility but it cannot be chosen too large because the system of equations becomes ill-conditioned with increasing penalty factor and leads to locking (trivial solution) whenever the penalty constraint takes a dominant role. The penalty based approach (also named as the "irreducible finite element flow formulation") is applied hereafter, such that the variational equation (3.2.1.11) is utilized.

3.2.2 Heat Transfer

The purpose of simulating heat transfer and heat generation is to model the effects of the temperature increase due to plastic work, to heat generated by electrical Joule heating and to temperature variation due to exchange of heat with the tools and the surrounding environment. In an arbitrary volume, the energy rate balance requires

$$\dot{q}_{in} - \dot{q}_{out} + \dot{q}_{generate} = \dot{q}_{store} \tag{3.2.2.1}$$

where $\dot{q}_{in}$ and $\dot{q}_{out}$ are the energy rates per unit volume into the volume and out of the volume, respectively. The heat rate per unit volume due to generation inside the volume is $\dot{q}_{generate}$, and $\dot{q}_{store}$ is the rate of stored energy per unit volume giving rise to a temperature gradient $\dot{T}$ according to

$$\dot{q}_{store} = \rho_m c_m \dot{T} \tag{3.2.2.2}$$

where ρ_m is the mass density and c_m is the heat capacity.

In the temperature range of melting and solidification, i.e. $T_{sol} < T < T_{liq}$ with solidus temperature T_{sol} and liquidus temperature T_{liq}, an effective heat capacity is defined to include an approximation of the latent heat L as follows [22],

$$\tilde{c}_m = c_m + \frac{L}{T_{liq} - T_{sol}} \tag{3.2.2.3}$$

Applying Fourier's law for heat conduction, $\dot{q} = -kT_{,i}$ with thermal conductivity k, and assuming the control volume to be infinitesimal, the transient heat diffusion equation can be obtained from (3.2.2.1) as

$$\left(kT_{,i}\right)_{,i} + \dot{q}_{generate} = \rho_m c_m \dot{T} \tag{3.2.2.4}$$

The heat generation has several contributions. In the material volume, heat generation exists due to dissipated energy from the plastic work and the electrical heat source due to Joule heating. On the boundary surface, the contributions are convection and radiation to the surroundings and to the tools as well as friction generated heat in contact interfaces with relative sliding.

The contribution from the plastic work, is the fraction of the plastic deformation energy dissipated as heat,

$$\dot{q}_{plastic} = \beta \sigma_{ij} \dot{\varepsilon}_{ij} = \beta \bar{\sigma} \dot{\bar{\varepsilon}} \quad (3.2.2.5)$$

where $\beta \approx 0.85 - 0.95$.

The generated Joule heating due to electrical resistance ρ and current density J is given by

$$\dot{q}_{electrical} = \rho J^2 \quad (3.2.2.6)$$

which will be analyzed more detailed in Sect. 3.2.3 dealing with the electrical field and resulting heat generation.

Newton's law for convection, applying to all free surfaces, is given by

$$\dot{q}_{convection} = h \left(T_s - T_f \right) \quad (3.2.2.7)$$

with heat transfer coefficient h, surface temperature T_s and temperature of the surroundings T_f.

Similarly Stefan-Boltzmann's law for radiation, applying to all free surfaces, is given by

$$\dot{q}_{radiation} = \varepsilon_{emis} \sigma_{SB} \left(T_s^4 - T_f^4 \right) \quad (3.2.2.8)$$

with emission coefficient ε_{emis} and Stefan-Boltzmann coefficient σ_{SB}.

At surfaces contacting the tools, convection follows

$$\dot{q}_{tool} = h_{lub} \left(T_s - T_{tool} \right) \quad (3.2.2.9)$$

where T_{tool} is the tool temperature and h_{lub} is the relevant convection coefficient, typically taken for an applied lubricant.

Finally, the heat generated by friction shear stresses τ_f in the contact interfaces with relative sliding v_r is given by

$$\dot{q}_{friction} = \tau_f |v_r| \quad (3.2.2.10)$$

The transient heat diffusion equation (3.2.2.4), was firstly implemented by Rebelo and Kobayashi [14, 15] in a finite element computer program for modeling thermo-mechanical metal forming processes, and subsequently implemented by Zhang et al. [22] for modeling the heat developed by Joule heating in resistance welding.

From this point of the presentation the thermal conductivity will be assumed constant within each integration domain, implying that $\left(k T_{,i} \right)_{,i}$ simplifies to $k T_{,ii}$. Under these circumstances and applying the classical Galerkin method, the heat transfer equation (3.2.2.4) can be written as follows,

$$\int_V k T_{,i} \delta T_{,i} \, dV + \int_V \rho_m c_m \dot{T} \delta T \, dV - \int_V \dot{q}_{generate} \delta T \, dV - \int_S k T_{,n} dS = 0$$

$$(3.2.2.11)$$

where $T_{,n}$ is the gradient of T along the outward normal to the surface S. The third term in (3.2.2.11) is the heat generated from plastic deformation (3.2.2.5) and Joule heating (3.2.2.6), and the fourth term is the heat flux on boundary surfaces. Along free surfaces S_{free} conduction and radiation follow (3.2.2.7) and (3.2.2.8), and along surfaces in contact with the tools S_{tool}, convection and friction generated heat follow (3.2.2.9) and (3.2.2.10). All these terms can be summarized as follows, .

$$\int_V kT_{,i}\delta T_{,i}\ dV + \int_V \rho_m\, c_m \dot{T}\delta T\ dV - \int_V \left(\dot{q}_{plastic} + \dot{q}_{electrical}\right)\delta T dV$$

$$+ \int_{S_{free}} \left(\dot{q}_{convection} + \dot{q}_{radiation}\right)dS + \int_{S_{tool}} \left(\dot{q}_{tool} - \dot{q}_{friction}\right)dS = 0$$

$$(3.2.2.12)$$

3.2.3 Electricity

The distribution of electric potential Φ utilized in the coupled electro-thermo-mechanical finite element implementation is based on Laplace's equation,

$$\Phi_{,ii} = 0 \qquad (3.2.3.1)$$

Although this approach considers the distribution of the electric potential to be solely determined by geometry under steady conditions ($\dot{\Phi} = 0$) [23], it is generally considered a good approach because an electric field has a much faster reaction rate than a temperature field.

Along boundaries with power supply, the electric potential is the supplied potential, $\Phi = \Phi_0$, and along free surfaces electric potential is zero. Integrating Laplace's equation for an arbitrary variation in the electric potential Φ and applying the divergence theorem, equation (3.2.3.1) becomes

$$\int_V \Phi_{,i}\delta\Phi_{,i}dV = \int_S \Phi_{,n}dS \qquad (3.2.3.2)$$

where $\Phi_{,n}$ is the normal gradient of the electric potential to the free surfaces. The right hand side of (3.2.3.2) can be omitted because $\Phi_{,n} = 0$ along free surfaces. Having solved the electric potential, the current density J in any direction is available through

$$J_i = \frac{1}{\rho}\Phi_{,i} \qquad (3.2.3.3)$$

Defining the squared current density as $J^2 = J_i J_i$, the heat generation rate due to Joule heating (3.2.2.6) is available through $\dot{q}_{electrical} = \rho J^2$.

3.3 Numerical Implementation

The above presented models for the mechanical, thermal and electrical responses can be combined and implemented in finite element computer programs based on the flow formulation. This section describes the coupling of the models and the details of computer implementation for each individual model.

3.3.1 Basic Coupling Procedures

Figure 3.1 includes a schematic outline of the couplings of the presented models. The thermal and mechanical models are generally coupled as shown in Fig. 3.1a for the purpose of modeling thermo-mechanical metal forming processes, whereas the electrical, thermal and mechanical models are coupled as shown in Fig. 3.1b for the electro-thermo-mechanical modeling of resistance welding processes.

Besides the immediate difference due to the electrical model, the two implementations differ by the number of times the mechanical model is applied during each step. In both cases the mechanical model is applied at the beginning of each step to setup a velocity field and a stress response.

The next step in the thermo-mechanical modeling of metal forming processes is to run the thermal model, and in this case it is run fully coupled with the mechanical model, such that the new temperature field and resulting changes in material

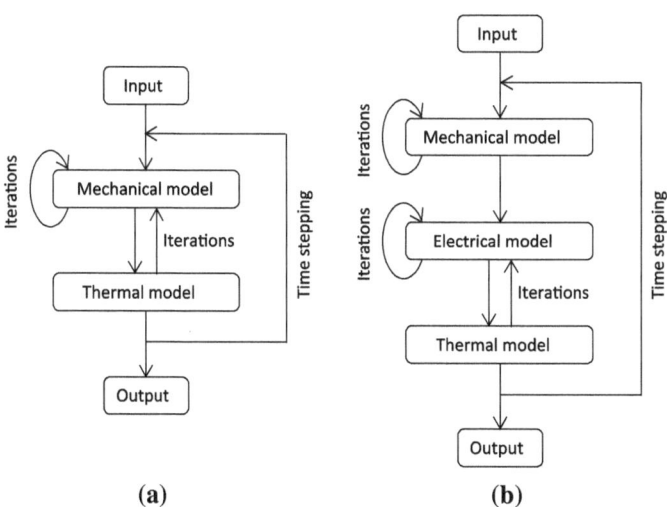

Fig. 3.1 Numerical coupling of mechanical, thermal and electrical models for **a** thermo-mechanical modeling of metal forming processes and **b** electro-thermo-mechanical modeling of resistance welding processes

properties are converged with the mechanical response including the heat generation at the end of each step.

When it comes to the electro-thermo-mechanical modeling of resistance welding, this strong coupling between the thermal and mechanical models is loosened
due to the very small time steps in order to capture the effects of the welding
process. For example, when using alternating current with frequency 50 Hz as
energy source, each half period has duration 10 ms, and proper modeling therefore requires time steps of 1 ms or preferably less. Instead of having a strong coupling, the implementation is relying on the small time steps in a weaker coupling,
where the new material properties of the resulting temperature is only affecting the
mechanical response from the following time step, and the corresponding change
in the heat generation due to plastic work is ignored due to the insignificant influence compared to the electrically generated heat.

In this type of implementation, the electrical model is applied after the mechanical model to supply the thermal model with the current density giving rise to the
heat generation. The electrical model is linear and thus inexpensive compared to
the mechanical model; hence the electrical and thermal models are strongly coupled such that the electrical model is run during each of the iterations of the thermal model. The implemented coupling is outlined in Fig. 3.1b and follows the
work of Zhang et al. [22].

3.3.2 Finite Elements

The discretization of the main equations dealing with the physics of plastic flow,
heat transfer and electricity is based on 8-node hexahedral elements under three-
dimensional conditions. Other elements could be employed in the discretization as
will be discussed in Chap. 5 in relation to mesh generation.

The 8-node hexahedral element provides three degrees of freedom in each
node for the velocity components of plastic flow in the mechanical model and one
degree of freedom for modeling the scalar fields of temperature and potential in
the thermal and electrical models, respectively.

In the mechanical model, discretization by hexahedral elements implies that
velocity inside an element is interpolated from its nodal values as follows,

$$\mathbf{u} = \mathbf{N}^T \mathbf{v} \tag{3.3.2.1a}$$

$$\mathbf{u} = \left\{ u_x, u_y, u_z \right\}^T \tag{3.3.2.1b}$$

$$\mathbf{v} = \left\{ u_x^1, u_y^1, u_z^1, u_x^2, u_y^2, u_z^2, \dots, u_z^8 \right\}^T \tag{3.3.2.1c}$$

where $\mathbf{u}$ is the vector containing the velocity components in an arbitrary location
within the element, $\mathbf{v}$ is the vector of nodal velocities and $\mathbf{N}$ is a matrix including
the shape functions N_i at the corresponding arbitrary location in natural coordinates ξ, η, ς (e.g. $N_i = (1 + \xi_i \xi)(1 + \eta_i \eta)(1 + \varsigma_i \varsigma)/8$).

The temperature and electric potential are interpolated similarly, except that the interpolation is for scalars rather than vectors of components. In all cases, the formulation is isoparametric, such that coordinates and field variables are interpolated by the same shape functions.

Matrix notation is introduced in what follows for better understanding the computer implementation of the discretized finite element equations.

3.3.3 Mechanical Model

3.3.3.1 Finite Element Discretization

The strain rate matrix $\mathbf{B}$ relating strain rates to nodal velocities is built from the derivatives of the shape functions in the following manner,

$$\dot{\boldsymbol{\varepsilon}} = \mathbf{B}\mathbf{v} = \mathbf{L}\mathbf{N}^T\mathbf{v}, \quad \mathbf{L} = \begin{bmatrix} \frac{\partial}{\partial x} & 0 & 0 \\ 0 & \frac{\partial}{\partial y} & 0 \\ 0 & 0 & \frac{\partial}{\partial z} \\ \frac{\partial}{\partial y} & \frac{\partial}{\partial x} & 0 \\ 0 & \frac{\partial}{\partial z} & \frac{\partial}{\partial y} \\ \frac{\partial}{\partial z} & 0 & \frac{\partial}{\partial x} \end{bmatrix} \tag{3.3.3.1}$$

Introducing a diagonal matrix $\mathbf{D} = \mathrm{diag}\left\{\frac{2}{3},\frac{2}{3},\frac{2}{3},\frac{1}{3},\frac{1}{3},\frac{1}{3}\right\}$ the effective strain rate (3.2.1.6) is written as

$$\left(\dot{\bar{\varepsilon}}\right)^2 = \dot{\boldsymbol{\varepsilon}}^T\mathbf{D}\dot{\boldsymbol{\varepsilon}} \tag{3.3.3.2}$$

or, in the following alternative matrix form after introducing (3.3.3.1) and defining $\mathbf{P} = \mathbf{B}^T\mathbf{D}\mathbf{B}$,

$$\left(\dot{\bar{\varepsilon}}\right)^2 = \mathbf{v}^T\mathbf{B}^T\mathbf{D}\mathbf{B}\mathbf{v} = \mathbf{v}^T\mathbf{P}\mathbf{v} \tag{3.3.3.3}$$

The volumetric strain rate $\dot{\varepsilon}_{ii}$ is expressed as follows,

$$\dot{\varepsilon}_{ii} = \mathbf{C}^T\mathbf{B}\mathbf{v} \tag{3.3.3.4}$$

with $\mathbf{C}$ being the vectorial form of the Kronecker delta δ_{ij}.

3.3.3.2 Newton–Raphson Iterative Procedure

By insertion of the above equations into (3.2.1.11), the first derivative of the energy rate functional is obtained as

$$\frac{\partial \Pi}{\partial \mathbf{v}} = \int_V \frac{\bar{\sigma}}{\dot{\bar{\varepsilon}}}\mathbf{P}\mathbf{v}dV + K\int_V \mathbf{B}^T\mathbf{C}\mathbf{C}^T\mathbf{B}\mathbf{v}dV - \int_S \mathbf{N}\mathbf{F}dS \tag{3.3.3.5}$$

where $\mathbf{F}$ is the matrix form of the applied boundary surface tractions $t_i = \sigma_{ij} n_j$.

The second derivative of the energy rate functional is obtained as

$$\frac{\partial^2 \Pi}{\partial \mathbf{v}^2} = \int_V \frac{\bar{\sigma}}{\dot{\bar{\varepsilon}}} \mathbf{P} dV + \int_V \left(\frac{1}{\dot{\bar{\varepsilon}}} \frac{\partial \bar{\sigma}}{\partial \dot{\bar{\varepsilon}}} - \frac{\bar{\sigma}}{\dot{\bar{\varepsilon}}^2} \right) \frac{1}{\dot{\bar{\varepsilon}}} \mathbf{P} \mathbf{v} \mathbf{v}^T \mathbf{P} dV + K \int_V \mathbf{B}^T \mathbf{C} \mathbf{C}^T \mathbf{B} dV$$

(3.3.3.6)

A second order linearization of (3.2.1.11) by Taylor expansion near an initial guess $\mathbf{v} = \mathbf{v}_0$ of the velocity field leads to

$$\underbrace{\left. \frac{\partial \Pi}{\partial \mathbf{v}} \right|_{\mathbf{v} = \mathbf{v}_0}}_{\equiv -\mathbf{f}} + \underbrace{\left. \frac{\partial^2 \Pi}{\partial \mathbf{v}^2} \right|_{\mathbf{v} = \mathbf{v}_0}}_{\equiv \mathbf{K}} \Delta \mathbf{v} = 0$$

(3.3.3.7)

which can be discretized by M finite elements and assembled to the system of equations

$$\sum_{m=1}^{M} \{ \mathbf{K} \Delta \mathbf{v} - \mathbf{f} \} = 0$$

(3.3.3.8)

From (3.3.3.7) and (3.3.3.8) it is seen that (3.3.3.6) is the stiffness matrix $\mathbf{K}$ and that (3.3.3.5) is the load vector $\mathbf{f}$ except for the sign. The stiffness matrix and the load vector are integrated in each element by Gauss integration and assembled into the global system of equations (3.3.3.8), which is solved for the velocity increment $\Delta \mathbf{v}$. The velocity $\mathbf{v}$ is updated according to

$$\mathbf{v}_n = \mathbf{v}_{n-1} + \alpha_{NR} \Delta \mathbf{v}_n$$

(3.3.3.9)

where n is the iteration number and $\alpha_{NR} \in \,]0; 1[$ is a deceleration coefficient to avoid overshooting and oscillations in the solution. The update is carried out until convergence,

$$\frac{|\Delta \mathbf{v}_n|}{|\mathbf{v}_{n-1}|} < \alpha_{conv}$$

(3.3.3.10)

that is, until the velocity field $\mathbf{v}$ is not changed considerable by including one more iteration. A typical value of α_{conv} is taken around 10^{-5}.

3.3.3.3 Direct Iterations

When applying direct iterations, the constitutive relation is evaluated at the previous converged velocity field, such that the iterations become linear. By insertion of (3.3.2.1a), (3.3.3.3) and (3.3.3.4) into the variation of the functional (3.2.1.11) and canceling out the virtual velocity field $\delta \mathbf{v}^T$ due to arbitrariness, the following system of equations is obtained,

$$\underbrace{\left(\int_V \frac{\bar{\sigma}}{\dot{\bar{\varepsilon}}} \mathbf{P} dV + K \int_V \mathbf{B}^T \mathbf{C} \mathbf{C}^T \mathbf{B} dV \right)}_{\equiv \mathbf{K}} \mathbf{v} = \underbrace{\int_S \mathbf{N} \mathbf{F} dS}_{\equiv \mathbf{f}}$$

(3.3.3.11)

where the stiffness matrix $\mathbf{K}$ and the load vector $\mathbf{f}$ are defined as well. Discretization by M finite elements and assembling into a global system of equations (3.3.3.11) lead to

$$\sum_{m=1}^{M} \{\mathbf{K}\mathbf{v} - \mathbf{f}\} = \mathbf{0} \qquad (3.3.3.12)$$

with update following

$$\mathbf{v}_n = \alpha_D \mathbf{v}_n + (1 - \alpha_D)\, \mathbf{v}_{n-1} \qquad (3.3.3.13)$$

In the above equation n is the iteration number and $\alpha_D \in \,]0;1[$ is a measure of the degree of updating, which acts as a stabilizer to avoid the solution to overshoot.

3.3.3.4 Combination of Direct and Newton–Raphson Iterative Procedures

The Newton–Raphson iterative procedure usually results in fast convergence near the actual solution, i.e. when a good estimate of the initial guess $\mathbf{v} = \mathbf{v}_0$ is provided. The initial velocity field can, however, be difficult to obtain and, therefore, the procedure employed in direct iterations is often applied to generate a velocity field close to the actual solution before the Newton–Raphson solution is applied for a fast convergence towards the required tolerance (3.3.3.10).

In the first step (that is, at the beginning of the numerical simulation), a velocity field corresponding to a constant strain rate in all elements may serve as the starting point for the direct iterations.

Schematic illustrations of the two iterative procedures are provided in Fig. 3.2 for a simplified one-dimensional velocity field. Figure 3.2a illustrates the fast convergence of the direct iterations in the early stages and Fig. 3.2b shows the fast convergence of the Newton–Raphson iterations near the solution. Figure 3.2c shows divergence with the Newton–Raphson iterative procedure in case of an initial guess for the velocity field further away from the actual solution or in case of a sudden complication due to non-linearities such as contact.

In case of divergence of the Newton–Raphson iterative procedure, convergence may be sought with direct iterations.

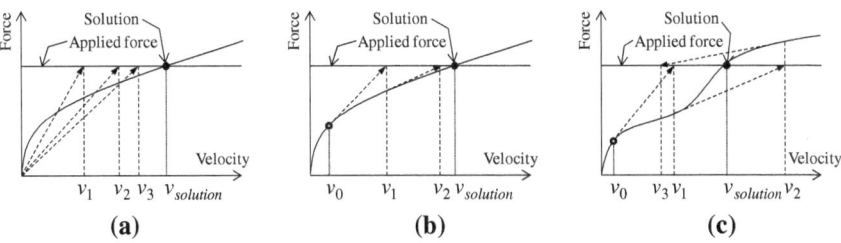

Fig. 3.2 Convergence schemes with subscript numbers referring to iteration number. **a** Direct iterations, **b** Newton–Raphson iterations with convergence, **c** Newton–Raphson iterations with divergence. *Subscript zero* identifies the initial guess for Newton–Raphson iterations

3.3.3.5 Selection of Deceleration Coefficients

As mentioned previously the deceleration coefficients α_D and α_{NR} control the degree of updating of both direct and Newton–Raphson iterative procedures. In case of direct iterations the selection of α_D is obtained after analyzing the ratios $\|\mathbf{v}_n - \mathbf{v}_{n-1}\| / \|\mathbf{v}_{n-1}\|$ and $\|\mathbf{R}_{n-1}\| / \|\mathbf{f}\|$ of the velocity $\mathbf{v}$ and residual $\mathbf{R}$ at iterations n and $n-1$, where

$$\mathbf{R}_{n-1} = \sum_{m=1}^{M} \{\mathbf{K}_{n-1}\mathbf{v}_n - \mathbf{f}\} \tag{3.3.3.14}$$

A similar approach is performed in case of Newton–Raphson iterative procedures, where the residual $\mathbf{R}$ is obtained from a Taylor expansion of the residual near the velocity estimate at the previous iteration,

$$\mathbf{R}(\mathbf{v}_n) \approx \mathbf{R}_n = \mathbf{R}_{n-1} + \left[\frac{\partial \mathbf{R}}{\partial \mathbf{v}}\right]_{n-1} \Delta\mathbf{v}_n = \mathbf{0} \tag{3.3.3.15}$$

In addition, it is also a good choice to determine the deceleration coefficient α_{NR} by means of a line search procedure that consider the residual $\mathbf{R}$ at the end of each iteration to be orthogonal to the velocity correction term $\Delta\mathbf{v}$ [24],

$$\Delta\mathbf{v}_n^T \cdot \mathbf{R}(\mathbf{v}_{n-1} + \alpha_{NR}\Delta\mathbf{v}_n) = \mathbf{0} \tag{3.3.3.16}$$

3.3.3.6 Domain Integration

The integration of the integrals in (3.3.3.5), (3.3.3.6) and (3.3.3.11) is performed by means of a selective Gauss integration scheme. Volume integrals are integrated by full integration (2^3 Gauss points) except for the second term in (3.3.3.5), the last term in (3.3.3.6) and the second term in (3.3.3.11), which are related to the volumetric part of the stiffness matrix $\mathbf{K}$. These terms are integrated by reduced Gauss integration (one Gauss point) to avoid locking. The surface integrals that include boundary pressure and friction along the tools are integrated by 5^2 Gauss points [25].

3.3.3.7 Stress Calculation

The direct results of (3.3.3.8) and (3.3.3.12) are the velocities and the strain rates. The strains are accumulated at the end of each simulation step by multiplying the strain rates by the increment of time and the effective strain allows determination of the effective stress directly from the applied material law.

The distribution of stress at the end of each simulation step requires determining the mean stress σ_m (refer to Sect. 3.2.1),

$$\sigma_m = \frac{K}{2}\dot{\varepsilon}_{kk} \tag{3.3.3.17}$$

and adding this value to the corresponding deviatoric stress obtained from the constitutive equations via the strain rate values (3.2.1.4) and (3.2.1.5),

$$\sigma_{ij} = \sigma'_{ij} + \delta_{ij}\sigma_m \tag{3.3.3.18}$$

The penalty K may be chosen as a constant value or as an adaptive value that changes for each element. If an adaptive value is chosen, small elements take larger penalty values because small elements are generally placed in the regions of more interest. The accuracy is thereby increased in the regions with refined mesh, while keeping the overall penalization as low as possible in order to diminish ill-conditioning of the matrix systems.

An option is to scale the penalty K according to the ratio of the maximum element volume to the actual element volume. If any scaling factor is above 10, all scaling factors are rescaled such that the maximum scaling is 10. This is to avoid very large penalty factors resulting in increased ill-conditioning.

3.3.3.8 Rigid Regions

To avoid singularities in the system of equations when having rigid regions, where the strain rates approach zero, a cut-off strain rate $\dot{\bar{\varepsilon}}_0$ is introduced [2]. Whenever $\dot{\bar{\varepsilon}} < \dot{\bar{\varepsilon}}_0$, the cut-off strain rate $\dot{\bar{\varepsilon}}_0$ replaces the actual strain rate to overcome the problem of singularities. The cut-off strain rate is taken as a value considerably smaller than the average strain rate of the deforming body. A too large value will model rigid regions poorly, and a too small value may lead to numerical inaccuracies.

An improvement of the above approach has been implemented by the authors to avoid excessive strain accumulation in rigid regions. The strain is only accumulated if the equivalent strain rate is increasing (which will not be the case if it is constantly equal to the cut-off strain rate) or if the equivalent strain has already exceeded a certain strain level meaning that the region should not be treated as rigid.

3.3.4 Thermal Model

Using the same shape functions as for the mechanical model, the temperature can be interpolated as

$$T = \mathbf{N}^T \mathbf{T} \tag{3.3.4.1}$$

where $\mathbf{T}$ contains the nodal temperatures and $\mathbf{N}$ contains the shape functions at positions to realize the summation over nodal values. Similarly a matrix $\mathbf{N}'$ is defined such that

$$T_{,i} = \mathbf{N}'^T \mathbf{T} \tag{3.3.4.2}$$

by having $N'_{ij} = N_{i,j}$.

Inserting (3.3.4.1) and (3.3.4.2) into (3.2.2.11) and canceling out the arbitrary temperature variation, the system of equations for the thermal model, $\mathbf{K}_c \mathbf{T} + \mathbf{C} \dot{\mathbf{T}} = \mathbf{q}$, becomes

$$
\underbrace{\int_V k \mathbf{N}' \mathbf{N}'^T dV}_{\equiv \mathbf{K}_c} \mathbf{T} + \underbrace{\int_V \rho_m c_m \mathbf{N} \mathbf{N}^T dV}_{\equiv \mathbf{C}} \dot{\mathbf{T}} = \underbrace{\int_V \dot{q}_{generate} \mathbf{N} dV + \int_S k T_{,n} \mathbf{N} dS}_{\equiv \mathbf{q}}
$$

$$(3.3.4.3)$$

where $\mathbf{K}_c$ is the heat conduction matrix, $\mathbf{C}$ is the heat capacity matrix and $\mathbf{q}$ includes the boundary flux and the source term. The right hand side $\mathbf{q}$ is expanded as follows to include the heat sources and heat loses due to Eqs. (3.2.2.5)–(3.2.2.10) as in (3.2.2.12),

$$
\mathbf{q} = \int_V \left(\dot{q}_{plastic} + \dot{q}_{electrical} \right) \mathbf{N} dV - \int_{S_{free}} \left(\dot{q}_{convection} + \dot{q}_{radiation} \right) \mathbf{N} dS
$$
$$
- \int_{S_{tool}} \left(\dot{q}_{tool} - \dot{q}_{friction} \right) \mathbf{N} dS
$$

$$(3.3.4.4)$$

The domain integration of the thermal system of equations (3.3.4.3) is performed over hexahedral elements in the usual manner, whereas the time integration is more complicated. The presence of the term including $\dot{\mathbf{T}}$ makes the system of equations differ from typical forms utilized in the mechanical models, e.g. (3.3.3.12). Details regarding the solution of the system of equations can be found in several references, e.g. in the pioneering work of Rebelo and Kobayashi [14, 15], which requires the utilization of the following time-stepping scheme,

$$
\mathbf{T}_{t+\Delta t} = \mathbf{T}_t + \Delta t \left[(1 - \theta) \dot{\mathbf{T}}_t + \theta \dot{\mathbf{T}}_{t+\Delta t} \right] \tag{3.3.4.5}
$$

where θ is a parameter varying between 0 and 1. A value of $\theta = 0.75$ is typically chosen.

3.3.5 Electrical Model

The shape functions and shape function derivatives are introduced similarly to (3.3.4.1) and (3.3.4.2), such that they interpolate the potential and its derivatives as follows,

$$
\Phi = \mathbf{N}^T \mathbf{\Phi} \tag{3.3.5.1}
$$

$$
\Phi_{,i} = \mathbf{N}'^T \mathbf{\Phi} \tag{3.3.5.2}
$$

Inserting (3.3.5.2) in (3.2.3.2) and canceling out the right hand side and the arbitrary potential variation, the discretized form of the electrical model (3.2.3.2) can be written as

$$\underbrace{\int_V \mathbf{N}' \mathbf{N}'^T dV}_{\equiv \mathbf{K}_e} \mathbf{\Phi} = \mathbf{0} \tag{3.3.5.3}$$

where $\mathbf{K}_e$ is the electrical conductance matrix to be integrated over elements and assembled into the global system of equations.

3.4 Incorporation of Anisotropy

Finite element modeling of manufacturing processes often treats materials as isotropic but when it comes to materials supplied as sheets, anisotropic behavior can be important due to the effect of prior rolling of the material. This section describes the implementation of Hill's quadratic anisotropic yield criterion [26] and Sect. 3.4.1 describes the necessary rotation between global axes and local material axes as they in general differ after deformation.

Hill's quadratic anisotropic yield criterion takes the following form,

$$f^a = \frac{1}{2} \frac{3}{2(F+G+H)} \left[F(\sigma_{22} - \sigma_{33})^2 + G(\sigma_{33} - \sigma_{11})^2 + H(\sigma_{11} - \sigma_{22})^2 \right. $$
$$\left. + 2L\sigma_{23}^2 + 2M\sigma_{31}^2 + 2N\sigma_{12}^2 \right] \tag{3.4.1}$$

where the anisotropic parameters, F, G, H, L, M and N, are to be determined from material testing through the following relations involving uniaxial and shear effective stresses $\bar{\sigma}_{ij}$,

$$\frac{1}{\bar{\sigma}_{11}^2} = G + H, \quad \frac{1}{\bar{\sigma}_{22}^2} = H + F, \quad \frac{1}{\bar{\sigma}_{33}^2} = F + G,$$
$$\frac{1}{\bar{\sigma}_{12}^2} = 2N, \quad \frac{1}{\bar{\sigma}_{23}^2} = 2L, \quad \frac{1}{\bar{\sigma}_{31}^2} = 2M \tag{3.4.2}$$

The yield function (3.4.1) can be written as

$$f^a = \frac{1}{2} \frac{3}{2(F+G+H)} \sigma_{ij} P_{ijkl} \sigma_{kl} \tag{3.4.3}$$

where

$$
P_{ijkl} = \begin{bmatrix}
P_{1111} & P_{1122} & P_{1133} & P_{1112} & P_{1123} & P_{1131} \\
P_{2211} & P_{2222} & P_{2233} & P_{2212} & P_{2223} & P_{2231} \\
P_{3311} & P_{3322} & P_{3333} & P_{3312} & P_{3323} & P_{3331} \\
P_{1211} & P_{1222} & P_{1233} & P_{1212} & P_{1223} & P_{1231} \\
P_{2311} & P_{2322} & P_{2333} & P_{2312} & P_{2323} & P_{2331} \\
P_{3111} & P_{3122} & P_{3133} & P_{3112} & P_{3123} & P_{3131}
\end{bmatrix}
$$

$$
= \begin{bmatrix}
G+H & -H & -G & 0 & 0 & 0 \\
-H & F+H & -F & 0 & 0 & 0 \\
-G & -F & F+G & 0 & 0 & 0 \\
0 & 0 & 0 & 2N & 0 & 0 \\
0 & 0 & 0 & 0 & 2L & 0 \\
0 & 0 & 0 & 0 & 0 & 2M
\end{bmatrix}
= \begin{bmatrix}
[\mathbf{P}^{UL}] & [\mathbf{0}] \\
[\mathbf{0}] & [\mathbf{P}^{LR}]
\end{bmatrix}
$$

$$(3.4.4)$$

The upper left (UL) and lower right (LR) submatrices are identified for later use.

The yield function f^a (3.4.3) is defined as counterpart to the isotropic yield function f associated with the von Mises yield criterion (3.2.1.3) and the effective stress is defined as

$$
\bar{\sigma} = \sqrt{\frac{3}{2\,(F+G+H)}} \left\{ \sigma_{ij}\, P_{ijkl}\, \sigma_{kl} \right\}^{\frac{1}{2}} \tag{3.4.5}
$$

which is the counterpart to the effective stress associated with von Mises isotropic yield criterion (3.2.1.7).

The proportionality factor associated with Hill's criterion is given by

$$
\dot{\lambda} = \frac{\dot{\bar{\varepsilon}}}{\bar{\sigma}} \tag{3.4.6}
$$

Insertion of (3.4.3) and (3.4.6) into the flow rule leads to the counterpart of the Levy–Mises constitutive equations; namely the relation between strain rates and deviatoric stresses that is consistent with Hill's criterion,

$$
\dot{\varepsilon}_{ij} = \frac{3}{2\,(F+G+H)} \frac{\dot{\bar{\varepsilon}}}{\bar{\sigma}} P_{ijkl}\sigma_{kl} = \frac{3}{2\,(F+G+H)} \frac{\dot{\bar{\varepsilon}}}{\bar{\sigma}} P_{ijkl}\sigma'_{kl} \tag{3.4.7}
$$

where the last equality is seen by insertion of $\sigma_{ij} = \sigma'_{ij} + \delta_{ij}\sigma_m$ and recognition of $P_{ijkl}\delta_{kl} = 0$ for any ij.

The deviatoric stress components are available through

$$
\sigma'_{ij} = \frac{2\,(F+G+H)\,\bar{\sigma}}{3} \frac{1}{\dot{\bar{\varepsilon}}} M_{ijkl}\dot{\varepsilon}_{kl} \tag{3.4.8}
$$

which follows from (3.4.7), except for the fact that P_{ijkl} is singular and therefore cannot be inverted. The tensor M_{ijkl} is therefore introduced instead of the non-existing inversion of P_{ijkl}. The structure of M_{ijkl} is

$$[\mathbf{M}] = \begin{bmatrix} \left[\mathbf{M}^{UL}\right] & [\mathbf{0}] \\ [\mathbf{0}] & \left[\mathbf{M}^{LR}\right] \end{bmatrix} \tag{3.4.9}$$

due to the structure of P_{ijkl}. The two non-zero submatrices are independent inversions of the corresponding submatrices defined in (3.4.4) as long as they would be regular. The lower right submatrix in (3.4.4) is regular, so $\left[\mathbf{M}^{LR}\right] = \left[\mathbf{P}^{LR}\right]^{-1}$. The upper left submatrix in (3.4.4) is singular, so $\left[\mathbf{P}^{UL}\right]^{-1}$ does not exist. Instead, $\left[\mathbf{M}^{UL}\right]$ is introduced such that

$$\left[\mathbf{M}^{UL}\right]\left[\mathbf{P}^{UL}\right]\left\{ \begin{matrix} \sigma'_{11} \\ \sigma'_{22} \\ \sigma'_{33} \end{matrix} \right\} = \frac{1}{3}\begin{bmatrix} 2 & -1 & -1 \\ -1 & 2 & -1 \\ -1 & -1 & 2 \end{bmatrix}\left\{ \begin{matrix} \sigma'_{11} \\ \sigma'_{22} \\ \sigma'_{33} \end{matrix} \right\} = \left\{ \begin{matrix} \sigma'_{11} \\ \sigma'_{22} \\ \sigma'_{33} \end{matrix} \right\} \tag{3.4.10}$$

since this will have the same effect as if $\left[\mathbf{M}^{UL}\right]$ was equal to $\left[\mathbf{P}^{UL}\right]^{-1}$. This is possible due to the last equality sign where it is utilized that $\sigma'_{ii} = 0$. The matrix $\left[\mathbf{M}^{UL}\right]$ satisfying (3.4.10) is written out together with $\left[\mathbf{M}^{LR}\right]$ to form the entire tensor M_{ijkl} with positions as defined in (3.4.4),

$$M_{ijkl} = \begin{bmatrix} Fk & -(F+G)k & -(H+F)k & 0 & 0 & 0 \\ -(F+G)k & Gk & -(G+H)k & 0 & 0 & 0 \\ -(H+F)k & -(G+H)k & Hk & 0 & 0 & 0 \\ 0 & 0 & 0 & \frac{1}{2N} & 0 & 0 \\ 0 & 0 & 0 & 0 & \frac{1}{2L} & 0 \\ 0 & 0 & 0 & 0 & 0 & \frac{1}{2M} \end{bmatrix},$$

$$k = \frac{1}{3(FG+FH+GH)}$$

$$\tag{3.4.11}$$

Insertion of (3.4.8) into $\bar{\sigma}\dot{\bar{\varepsilon}} = \sigma_{ij}\dot{\varepsilon}_{ij}$ yields an expression for the effective strain rate similar to that of the isotropic formulation based on von Mises' yield criterion (3.2.1.6),

$$\dot{\bar{\varepsilon}} = \sqrt{\frac{2(F+G+H)}{3}}\left\{\dot{\varepsilon}_{ij}M_{ijkl}\dot{\varepsilon}_{kl}\right\}^{\frac{1}{2}} \tag{3.4.12}$$

which by insertion of (3.4.11) and utilization of $\dot{\varepsilon}_{ii} = 0$ leads to

$$\dot{\bar{\varepsilon}} = \sqrt{\frac{2(F+G+H)}{3}}\left\{ \frac{F\dot{\varepsilon}_{11}^2 + G\dot{\varepsilon}_{22}^2 + H\dot{\varepsilon}_{33}^2}{FG+FH+GH} \right.$$

$$\left. + \frac{(2\dot{\varepsilon}_{12})^2}{2N} + \frac{(2\dot{\varepsilon}_{23})^2}{2L} + \frac{(2\dot{\varepsilon}_{31})^2}{2M} \right\}^{\frac{1}{2}} \tag{3.4.13}$$

From (3.4.13) it is seen that an equation similar to (3.3.3.2) can be set up by defining another diagonal **D**-matrix, namely

$$\mathbf{D}^a = \frac{2\,(F+G+H)}{3}\,\mathrm{diag}\left\{\frac{\{F,G,H\}}{FG+FH+GH},\frac{1}{2N},\frac{1}{2L},\frac{1}{2M}\right\} \quad (3.4.14)$$

which is related to the effective strain rate like in (3.3.3.2). The anisotropic finite element formulation follows the derivations in Sect. 3.3 with substitution of (3.4.14) into (3.3.3.2) and (3.3.3.3).

3.4.1 Rotation Between Global Axes and Material Axes

In the above formulation, $\mathbf{D}^a$ refers to the global coordinate system, which may not be the same as the material coordinate system. In general, part of the deformation is rigid body rotation, which gives rise to misalignment between material axes and global axes. Therefore an incremental rotation matrix is set up to rotate $\mathbf{D}^a$ in each step according to the rigid body rotations associated with the previous step.

From the updated nodal velocities, the spin rate tensor,

$$\dot{\omega}_{ij} = \frac{1}{2}\left(\frac{\partial u_i}{\partial x_j} - \frac{\partial u_j}{\partial x_i}\right) = \begin{bmatrix} 0 & \omega_{12} & \omega_{13} \\ -\omega_{12} & 0 & \omega_{23} \\ -\omega_{13} & -\omega_{23} & 0 \end{bmatrix} \quad (3.4.1.1)$$

can be calculated in each step in each element. It is set up for the central point (in natural coordinates) of each element through the shape function derivatives. Assuming small incremental rigid body rotations, the incremental rotation matrix is approximated by adding the unit matrix and the incremental spin matrix, i.e.

$$\Delta\mathbf{R} = \mathbf{I} + \omega\Delta t \quad (3.4.1.2)$$

This incremental rotation matrix is a 3×3 matrix, which rotates another 3×3 matrix $\tilde{\mathbf{D}}^a$ through

$$\tilde{\mathbf{D}}^a_k = \Delta\mathbf{R}\tilde{\mathbf{D}}^a_{k-1}\Delta\mathbf{R}^T \quad (3.4.1.3)$$

where k represents step number. $\mathbf{D}^a$ in (3.4.14) is transferred into a 3×3 format during the rotation by the following translations between positions, pos_i (used towards right before rotation and used towards left after rotation),

$$\mathrm{diag}\,\{pos_1, pos_2, pos_3, pos_4, pos_5, pos_6\} \leftrightarrow \begin{bmatrix} pos_1 & pos_4 & pos_6 \\ pos_4 & pos_2 & pos_5 \\ pos_6 & pos_5 & pos_3 \end{bmatrix}$$

$$(3.4.1.4)$$

This translation follows the translation between the stress vector and the stress matrix defined as

$$\boldsymbol{\sigma} = \{\sigma_{11}, \sigma_{22}, \sigma_{33}, \sigma_{12}, \sigma_{23}, \sigma_{31}\}^T \leftrightarrow \begin{bmatrix} \sigma_{11} & \sigma_{12} & \sigma_{31} \\ \sigma_{12} & \sigma_{22} & \sigma_{23} \\ \sigma_{31} & \sigma_{23} & \sigma_{33} \end{bmatrix} \qquad (3.4.1.5)$$

The stress calculation should also be carried out with attention to rotation. The deviatoric stress is related to the strain rate according to (3.4.8). In this equation, M_{ijkl} is defined in a material coordinate system, whereas the available strain rates are defined relative to the global coordinate system. Due to possible material rotation, these two systems may not coincide. It is therefore necessary to rotate one of the quantities from one system to the other. The implemented procedure is as follows; the strain rate in each element is rotated from global axes to material axes through the following rotation,

$$\dot{\boldsymbol{\varepsilon}}_{mat} = \mathbf{R} \dot{\boldsymbol{\varepsilon}} \mathbf{R}^T \qquad (3.4.1.6)$$

where $\mathbf{R}$ is the accumulated rotation during N_{rot} rotation steps defined as

$$\mathbf{R} = \mathbf{I} + \sum_{i=1}^{N_{rot}} (\omega \Delta t)_i \qquad (3.4.1.7)$$

Now in material axes, the deviatoric stress can be computed by insertion of (3.4.1.6) into (3.4.8), and finally the deviatoric stress in the global system is available by rotation,

$$\boldsymbol{\sigma}' = \mathbf{R}^T \boldsymbol{\sigma}'_{mat} \mathbf{R} \qquad (3.4.1.8)$$

Rotation of the Cauchy stress is possible without introduction of artificial contributions from rigid body rotation since it is an objective stress measure. The remaining stress calculation follows (3.3.3.17) and (3.3.3.18).

As a final remark to the anisotropic formulation, it should be mentioned that with $F = G = H = 1$ and $L = M = N = 3$, the anisotropic formulation reduces to the isotropic formulation.

3.5   Incorporation of Elastic Effects

The core of the mechanical model is the rigid-plastic/viscoplastic flow formulation as presented so far. In this formulation the elastic effects are neglected due to the large deformations typically simulated. The elastic effects may, however, be of importance in some cases when only part of the volume is heavily deformed. In these cases, the remaining volume will only deform slightly, such that the elastic part should not be neglected.

An example where elastic deformation is of importance is resistance welding including bending of a sheet (gap between sheets or welding of a component to a sheet structure). In this case the overall deformation is governed by elastic deformation and only local deformation is governed by plasticity. The amount of elastic bending can be of importance to the actual contact area, which is essential for the welding process.

Elastic effects can be included in computer programs based on the finite element flow formulation following the procedure or variants of the procedure proposed by Mori et al. [27]. By doing this, the elastic effects are captured while the advantages of the flow formulation are kept for the remaining elements considered rigid-plastic due to the large deformations. A possible implementation of this procedure can be implemented as described in what follows.

All elements are initialized as elastic elements before loading. After loading to the vicinity of the yield stress Y, the relevant elements are turned into elastoplastic elements, and after further loading the relevant elements are turned into rigid-plastic elements ignoring any further elastic deformation. In order for the programs to be more efficient, a range of stress is assigned to define the elastoplastic behavior of the elements. With reference to Fig. 3.3a, the constitutive laws are applied as follows,

$$\text{constitutive law} = \begin{cases} \text{elastic,} & \bar{\sigma} \leq f_l Y \\ \text{elastoplastic,} & f_l Y < \bar{\sigma} < f_u Y \\ \text{rigid-plastic/viscoplastic,} & \bar{\sigma} \geq f_u Y \end{cases}$$

(3.5.1)

where typical factors are chosen around $f_l = 0.95$ and $f_u = 1.01$ where the flattened curve after yielding is reflected in the upper factor being closer to unity than the lower factor.

A stress situation in the vicinity of yielding is illustrated in Fig. 3.3b, where a stress path is exceeding the yield stress of the material causing strain hardening. The present stress state P is elastic with effective stress less than the yield stress, $\bar{\sigma}_t < Y$. The assumed load increment will cause a stress path through yielding (point Q) followed by strain hardening to a stress state in point R with effective stress, $\bar{\sigma}_{t+\Delta t} = \bar{\sigma}_t + \Delta\bar{\sigma}_{t+\Delta t}$, equal to the new flow stress. A ratio R_e of the elastic part of the stress to the total stress increment is defined and approximated, respectively, as follows with reference to Fig. 3.3b,

$$R_e = \frac{\overline{PQ}}{\overline{PR}} \approx \frac{\overline{PW}}{\overline{PS}} = \frac{Y - \bar{\sigma}_t}{\bar{\sigma}_{t+\Delta t} - \bar{\sigma}_t}$$

(3.5.2)

Yamada et al. [28] presented the correct solution corresponding to $\frac{\overline{PQ}}{\overline{PR}}$, but the approximation by $\frac{\overline{PW}}{\overline{PS}}$ is considered sufficient for the present purpose.

The ratio R_e was originally used to scale the load increment according to the elastic element closest to yielding to achieve a situation where it just reaches the yield stress. Hereafter, the element will be considered plastic. Another approach is

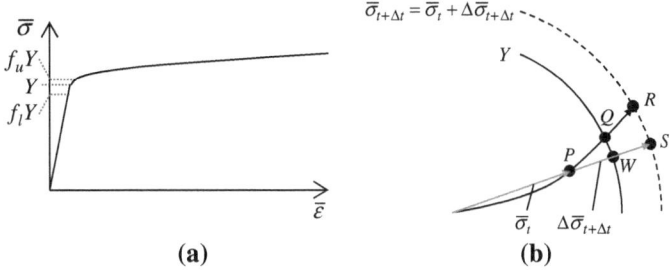

Fig. 3.3 Definitions in the vicinity of the yield stress Y. **a** Limits defining elastic, elastoplastic and rigid-plastic/viscoplastic regions. **b** Stress path for definition of elastic and elastoplastic fractions of stress increment

to avoid splitting the time step (corresponding to the load increment). The ratio of the elastic contribution to the stress increment is instead used to scale the amount of the stress–strain matrix stemming from either the elastic relation or the elastoplastic relation according to

$$\Delta\boldsymbol{\sigma} = \mathbf{D}^{ep}\Delta\boldsymbol{\varepsilon} = \left(R_e\mathbf{D}^e + (1 - R_e)\mathbf{D}^p\right)\Delta\boldsymbol{\varepsilon} \qquad (3.5.3)$$

where e refers to elasticity and p to plasticity.

For pure elasticity (3.5.3) reduces to Hooke's generalized law after inversion and the elastic stress–strain matrix is written as follows,

$$\mathbf{D}^e = \frac{E}{(1+v)(1-2v)}\begin{bmatrix} 1-v & v & v & 0 & 0 & 0 \\ v & 1-v & v & 0 & 0 & 0 \\ v & v & 1-v & 0 & 0 & 0 \\ 0 & 0 & 0 & \frac{1-2v}{2} & 0 & 0 \\ 0 & 0 & 0 & 0 & \frac{1-2v}{2} & 0 \\ 0 & 0 & 0 & 0 & 0 & \frac{1-2v}{2} \end{bmatrix}$$

$$(3.5.4)$$

For elastoplasticity, (3.5.3) resembles the inverse Prandtl-Reuss equations. The starting point is taken by the deviatoric part of the Prandtl-Reuss equations,

$$d\varepsilon'_{ij} = \sigma'_{ij}d\lambda + \frac{d\sigma'_{ij}}{2G} \qquad (3.5.5a)$$

$$d\lambda = \frac{3}{2}\frac{d\bar{\varepsilon}^p}{\bar{\sigma}} = \frac{3}{2}\frac{d\bar{\sigma}}{\bar{\sigma}H'} \qquad (3.5.5b)$$

where $G = \frac{E}{2(1+v)}$ is the shear modulus and $H' = \frac{d\bar{\sigma}}{d\bar{\varepsilon}^p}$ is the slope of the stress–strain curve. The corresponding elastoplastic stress–strain matrix originally obtained by Yamada et al. [28] is built by inverting (3.5.5a) and can be written as follows,

$$
\mathbf{D}^P = \frac{1}{1+v}
\begin{bmatrix}
\frac{1-v}{1-2v} - \frac{\sigma_{11}'^2}{S} & \frac{v}{1-2v} - \frac{\sigma_{11}'\sigma_{22}'}{S} & \frac{v}{1-2v} - \frac{\sigma_{11}'\sigma_{33}'}{S} & -\frac{\sigma_{11}'\sigma_{12}}{S} & -\frac{\sigma_{11}'\sigma_{23}}{S} & -\frac{\sigma_{11}'\sigma_{31}}{S} \\[6pt]
\frac{v}{1-2v} - \frac{\sigma_{11}'\sigma_{22}'}{S} & \frac{1-v}{1-2v} - \frac{\sigma_{22}'^2}{S} & \frac{v}{1-2v} - \frac{\sigma_{22}'\sigma_{33}'}{S} & -\frac{\sigma_{22}'\sigma_{12}}{S} & -\frac{\sigma_{22}'\sigma_{23}}{S} & -\frac{\sigma_{22}'\sigma_{31}}{S} \\[6pt]
\frac{v}{1-2v} - \frac{\sigma_{11}'\sigma_{33}'}{S} & \frac{v}{1-2v} - \frac{\sigma_{22}'\sigma_{33}'}{S} & \frac{1-v}{1-2v} - \frac{\sigma_{33}'^2}{S} & -\frac{\sigma_{33}'\sigma_{12}}{S} & -\frac{\sigma_{33}'\sigma_{23}}{S} & -\frac{\sigma_{33}'\sigma_{31}}{S} \\[6pt]
-\frac{\sigma_{11}'\sigma_{12}}{S} & -\frac{\sigma_{22}'\sigma_{12}}{S} & -\frac{\sigma_{33}'\sigma_{12}}{S} & \frac{1}{2} - \frac{\sigma_{12}'^2}{S} & -\frac{\sigma_{12}'\sigma_{23}}{S} & -\frac{\sigma_{12}'\sigma_{31}}{S} \\[6pt]
-\frac{\sigma_{11}'\sigma_{23}}{S} & -\frac{\sigma_{22}'\sigma_{23}}{S} & -\frac{\sigma_{33}'\sigma_{23}}{S} & -\frac{\sigma_{12}'\sigma_{23}}{S} & \frac{1}{2} - \frac{\sigma_{23}'^2}{S} & -\frac{\sigma_{23}'\sigma_{31}}{S} \\[6pt]
-\frac{\sigma_{11}'\sigma_{31}}{S} & -\frac{\sigma_{22}'\sigma_{31}}{S} & -\frac{\sigma_{33}'\sigma_{31}}{S} & -\frac{\sigma_{12}'\sigma_{31}}{S} & -\frac{\sigma_{23}'\sigma_{31}}{S} & \frac{1}{2} - \frac{\sigma_{31}'^2}{S}
\end{bmatrix}
$$

$$\tag{3.5.6}$$

with

$$
S = \frac{2}{3}\bar{\sigma}^2 \left(1 + \frac{1}{3G}\frac{d\bar{\sigma}}{d\bar{\varepsilon}^p} \right)
\tag{3.5.7}
$$

The elastoplastic solution presented by (3.5.3) with elastic and elastoplastic stress relations by (3.5.4) and (3.5.6) requires the stress to be incremented in each step, which is not the case in the flow formulation where the stress is given solely by the accumulated effective strain and the strain rate of the current step. In the flow formulation, the stress is therefore not necessarily saved between steps unless written to result files. On the contrary, in the solid formulations, the stress field of the previous step is of importance as the new step is only solving a stress increment. The stress of the previous step enters the equations as an initial stress, and in the end of the step it is incremented by the solution obtained in (3.5.3).

In general, the deformation will include rigid body motion between simulation steps. It is therefore necessary at each step to rotate the stress from the previous step into the new configuration, both for the role of initial stress and for the incremental update in the end of the step. With incremental rotation as defined in (3.4.1.2) and calculated stress increment $\Delta\sigma_{t+\Delta t}$, the stress after the new time step is

$$
\sigma_{t+\Delta t} = \Delta\sigma_{t+\Delta t} + \Delta\mathbf{R}\sigma_t\Delta\mathbf{R}^T
\tag{3.5.8}
$$

where the last term is identical to the stress field of the previous time step rotated into the new configuration. This term is also applied as the initial stress.

The presented formulation includes a mixture of elastic, elastoplastic and rigid-plastic/viscoplastic elements. Whenever elastic effects are relevant, all the elements are initialized as elastic as mentioned previously. They are changed to elastoplastic elements in the vicinity of yielding according to (3.5.1) and later changed to rigid-plastic/viscoplastic elements. The different states of the elements are working simultaneously, implying that typical situations will include a local deformation zone with rigid-plastic/viscoplastic elements, a transition zone of elastoplastic elements while the remaining elements are elastic.

Elastic unloading at the end of a simulation is performed by changing all elements to the elastic state and performing one more iteration step with the actual stress field as the initial stress. Dynamic elastic unloading was covered by Mori et al. [27] by changing elements back to the elastic state according to (3.5.1) and Fig. 3.3a.

References

1. Boer CR, Gudmundson P, Rebelo N (1982) Comparison of elastoplastic FEM, rigid-plastic FEM and experiments for cylinder upsetting. In: Pittman JFT (ed) Numerical methods for industrial forming processes. Pineridge Press, Swansea
2. Kobayashi S, Oh SI, Altan T (1989) Metal forming and the finite element method. Oxford University Press, Oxford
3. Brännberg N, Mackerle J (1994) Finite element methods and material processing technology. Eng Comput 11:413–455
4. Mackerle J (1998) Finite element methods and material processing technology, an addendum (1994–1996). Eng Comput 15:616–690
5. Mackerle J (2004) Finite element analyses and simulations of sheets metal forming processes. Eng Comput 21:891–940
6. Singh S (2004) Can simulation of the welding process help advance the state of the art in resistance welding? In: Proceedings of the 3rd international seminar on advances in resistance welding, Berlin, Germany, pp 5–11
7. Lee CH, Kobayashi S (1973) New solutions to rigid plastic deformation problems using a matrix method. J Eng Ind, ASME 95:865–873
8. Cornfield GC, Johnson RH (1973) Theoretical prediction of plastic flow in hot rolling including the effect of various temperature distributions. J Iron Steel Inst 211:567–573
9. Zienkiewicz OC, Godbole PN (1974) Flow of plastic and viscoplastic solids with special reference to extrusion and forming processes. Int J Numer Meth Eng 8:3–16
10. Altan T, Knoerr M (1992) Application of 2D finite element method to simulation of cold forging process. J Mater Process Technol 35:275–302
11. Zienkiewicz OC, Jain PC, Oñate E (1978) Flow of solids during forming and extrusion. Some aspects of numerical solutions. Int J Solids Struct 12:15–38
12. Zienkiewicz OC, Oñate E, Heinrich JC (1978) Plastic flow in metal forming—I. Coupled thermal behavior in extrusion—II. Thin sheet forming. In: Proceedings of winter annual meeting of ASME on application of numerical methods to forming processes, San Francisco, vol 28, p 107
13. Zienkiewicz OC, Oñate E, Heinrich JC (1981) A general formulation for coupled thermo flow of metals using finite elements. Int J Numer Meth Eng 17:1497–1514
14. Rebelo N, Kobayashi S (1980) A coupled analysis of viscoplastic deformation and heat transfer—I. Theoretical considerations. Int J Mech Sci 22:699–706
15. Rebelo N, Kobayashi S (1980) A coupled analysis of viscoplastic deformation and heat transfer—II. Applications. Int J Mech Sci 22:707–718
16. Nied HA (1984) The finite element modeling of the resistance spot welding process. Welding Res Suppl 63:123–132
17. Zhu W-F, Lin ZQ, Lai X-M, Luo A-H (2006) Numerical analysis of projection welding on auto-body sheet metal using a coupled finite element method. Int J Adv Manuf Technol 28:45–52
18. Zhang W (2010) Weld planning with optimal welding parameters by computer simulations and optimizations. In: Proceedings of the 6th international seminar on advances in resistance welding, Hamburg, Germany, pp 119–127

19. Nielsen C, Martins PAF, Zhang W, Bay N (2011) Mechanical contact experiments and simulations. Steel Res Int 82:645–650
20. Tekkaya AE, Martins PAF (2009) Accuracy, reliability and validity of finite element analysis in metal forming: a user's perspective. Eng Comput 26:1026–1055
21. Alves ML, Rodrigues JMC, Martins PAF (2004) Three-dimensional modelling of forging processes by the finite element flow formulation. J Eng Manuf 218:1695–1707
22. Zhang W, Jensen HH, Bay N (1997) Finite element modeling of spot welding similar and dissimilar metals. In: Proceedings of the 7th international conference on computer technology in welding, San Francisco, USA, pp 364–373
23. Greenwood JA, Williamson JBP (1958) Electrical conduction in solids II. Theory of temperature-dependent conductors. Proc R Soc Lond A, Math Phy Sci 246:13–31
24. Alves ML (2004) Modelação numérica e análise experimental de operações de forjamento. PhD thesis, IST-Technical University of Lisbon (in Portuguese)
25. Barata Marques MJM, Martins PAF (1990) Three-dimensional finite element contact algorithm for metal forming. Int J Numer Meth Eng 30:1341–1354
26. Hill R (1950) The mathematical theory of plasticity. Oxford University Press, London
27. Mori K, Wang CC, Osakada K (1996) Inclusion of elastic deformation in rigid-plastic finite element analysis. Int J Mech Sci 38:621–631
28. Yamada Y, Yoshimura N, Sakurai T (1968) Plastic stress-strain matrix and its application for the solution of elastic-plastic problems by the finite element method. Int J Mech Sci 10:343–354

Chapter 4
Contact Modeling

Due to the highly non-linear behavior, contact modeling remains among the more difficult disciplines within finite element simulations. Contact between workpieces and tooling and in-between workpieces defines the shape of formed components in metal forming as well as the contact conditions in resistance welding between the components to be joined and the welding electrodes. Section 4.1 presents a direct contact algorithm to handle the contact between a deformable workpiece and rigid tools. Section 4.2 presents a variational approach to the contact between deformable objects and Sect. 4.3 presents an industrial application by fabrication of seamless cylindrical reservoirs by tube forming that combines the two aforementioned contact modeling approaches.

Descriptions are given based on mechanical contact while thermal and electrical contacts are included by simplification of the mechanical description. The mechanical contact conditions can be separated into normal constraints and tangential constraints. The normal constraint is always that the contacting surfaces cannot penetrate into each other. The tangential constraints depend on the treatment of friction. In case of a frictionless approach, there are no tangential constraints and in case of full sticking, the tangential constraints are similar to the normal constraint since relative sliding is not allowed. In case of frictional conditions (including combined sticking and sliding), the constraints are governed by the employed friction law. At low, medium and high normal pressures, the following three friction laws are commonly employed:

- Amonton-Coulomb $\tau_f = \mu p$, typically assumed for normal pressure below $\sim 1.5\, p/Y$.
- Law of constant friction (Tresca) $\tau_f = mk$, typically assumed above $\sim 3\, p/Y$.
- Wanheim-Bay general friction model $\tau_f = f\alpha k$, applicable over the entire range of normal pressure and especially relevant in the range between the two aforementioned models above $\sim 1.5\, p/Y$ and below $\sim 3\, p/Y$.

In the above friction models, the friction shear stress is τ_f, p is the normal pressure and Y is the material flow stress of the softest contact surface, k is the shear flow

C. V. Nielsen et al., *Modeling of Thermo-Electro-Mechanical Manufacturing Processes*, SpringerBriefs in Applied Sciences and Technology, DOI: 10.1007/978-1-4471-4643-8_4, © The Author(s) 2013

stress, μ is the friction coefficient, f and m are the friction factors and α is the ratio of the real contact area to the nominal contact area.

4.1 Contact Between Workpiece and Tooling

During non-stationary processes, boundary conditions are progressively modified as a result of the interaction between workpiece and tooling. The contact algorithm implemented in both I-Form3 and SORPAS 3D is based on Barata Marques and Martins [1] and requires the workpieces to be discretized by hexahedral elements and tools (treated as rigid) to be discretized by spatial triangular surface elements. This discretization of tool surfaces had originally been proposed by Chenot [2], while Shiau and Kobayashi [3] and Yoon and Yang [4] preferred to describe the tool geometry by Bezier surfaces. However, the choice of a discretization by spatial triangles is somewhat natural in finite element modeling which is already based on discretization procedures.

The resulting contact formulation is based on node-to-triangle contact as illustrated in Fig. 4.1 by a workpiece node contacting a triangular element of the tool. Boundary nodes, like N_P in Fig. 4.1, are analyzed for each triangular surface element of the tool. The orthogonal projection N_{P_*} of node N_P to the plane spanned by the triangle is calculated. Figure 4.1b shows an example of the orthogonal projection being inside the considered triangular element, which is one of the conditions for being in contact. Figure 4.1c shows an example of the orthogonal projection lying outside, and hence node N_P and this triangular element are not in contact. The evaluation of whether or not the projection lies inside the triangle is based on a comparison of the total area of the triangle $A_{N_1 N_2 N_3}$ and the area sum

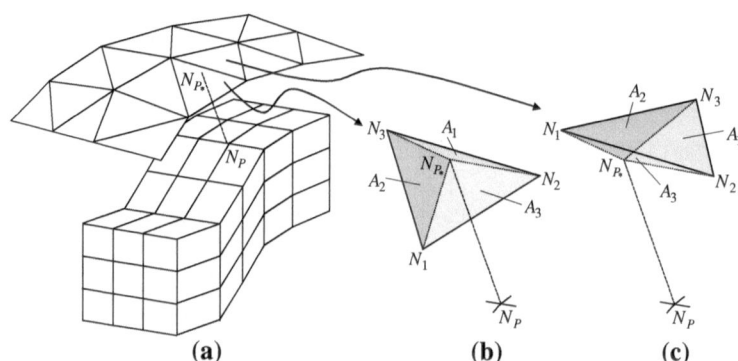

Fig. 4.1 Contact between the hexahedral mesh and the triangular surface mesh of a rigid tool. **a** Node N_P and its projection N_{P_*} in a triangular element, **b** normal projection N_{P_*} of node N_P lying inside triangular surface element and **c** normal projection N_{P_*} of node N_P lying outside triangular surface element

$A_1 + A_2 + A_3$ of the triangles spanned by the projection point and two of the triangle vertices. If the point is inside, the two areas are identical. Another condition for being in contact is that the distance between N_P and N_{P_*} is less than a specified value in order to avoid nodes far from the tool to be considered in contact.

The time increment necessary for a nodal point to get in contact with the tools is evaluated implicitly ($\theta = 0$) or explicitly ($\theta = 1$) according to

$$\mathbf{x}_{t+\Delta t} = \mathbf{x}_t + \left[\theta\,\mathbf{u}_t + (1-\theta)\,\mathbf{u}_{t+\Delta t}\right]\Delta t \qquad (4.1.1)$$

The implemented computer program is calculating the time increment based on the explicit approach, such that the time needed for each of the potential nodes to get in contact with a tool is calculated according to

$$\Delta t_p = \frac{\overline{N_P N_{P_*}}}{v_{N_P} - v_{N_{P_*}}} \qquad (4.1.2)$$

where the denominator is the normal velocity difference between the candidate node N_P and its projection N_{P_*} on the tool, which if it is negative corresponds to an increasing gap and in that case it is discarded as a candidate. Among the candidates, the minimum time $\Delta t_P^{\min}$ from (4.1.2) is decisive for the following time increment. If the time step is larger than the minimum time for a contact point to arise, it is split to $\Delta t = \Delta t_P^{\min}$. All points getting in contact to the tools within a specified tolerance in the following step are projected to the tool and assigned boundary conditions to enforce the points to follow the movement of the tool.

Taking the constant friction law as an example, the friction stress $\tau_f = mk$ acts in the opposite direction of the relative velocity $\mathbf{u}_r$ between workpiece material and tool and can therefore be written as

$$\tau_f = -mk\,\frac{\mathbf{u}_r}{|\mathbf{u}_r|} \qquad (4.1.3)$$

This friction model is illustrated in Fig. 4.2a at the vicinity of a neutral point (no relative velocity). The derivative of the friction stress with respect to the relative velocity is also shown as it is relevant for the finite element implementation, and it is seen that the derivative goes to infinity. To avoid this singularity, Chen and Kobayashi [5] proposed the following approximation,

$$\tau_f \cong -\frac{2}{\pi}\,mk \cdot \arctan\left(\frac{\mathbf{u}_r}{u_0}\right) \qquad (4.1.4)$$

which resembles the friction stress as shown in Fig. 4.2b when u_0 is a constant much smaller than the magnitude of the relative sliding velocity. The friction contribution to the functional Π (3.2.1.8) is

$$\Pi_f = \int_{S_{tool}}\left[\int_0^{|u_r|}\tau_f\,d\mathbf{u}_r\right]dS \qquad (4.1.5)$$

Fig. 4.2 Friction between
workpiece and rigid tools.
a Relative velocity (*upper*),
corresponding friction stress
according to the constant
friction law (*middle*) and
the derivative of the friction
stress with respect to the
relative velocity. **b** Modified
friction stress according to
(4.1.4) and its derivative

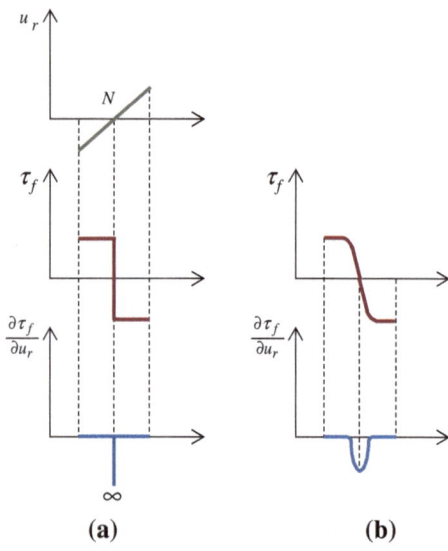

The first and second variations of this term are evaluated and added to equations
(3.3.3.5) and (3.3.3.6), thereby entering (3.3.3.7) and (3.3.3.11) also. The derivatives
of (4.1.5) are integrated by 5×5 Gauss quadrature [1].

Once a node is in contact with the tools, it is kept in contact until the normal
stress eventually becomes positive, which corresponds to a release of contact.
Whenever a node is in contact it is treated mechanically as above, but also thermal
and electrical effects may be relevant. The thermal effects are due to heat exchange
with the tool (3.2.2.9) and friction generated heat (3.2.2.10). The electrical boundary
conditions are either an applied potential or isolation (isolation is similar to a free
surface).

4.2 Contact Between Deformable Bodies

An indirect, variational approach is taken to the modeling of contact between
deformable objects. A modification to the variation of the functional expressing
the total energy-rate of the system is performed by adding a term due to the contact constraints. Traditionally, Lagrange multipliers or the penalty method has
been applied. The method of Lagrange multipliers solves the problem exactly, but
at the cost of additional unknowns.

The penalty method does not include additional unknowns, but suffers from
a compromise in choosing high penalty factors for improving accuracy and low
penalty factors for avoiding ill-conditioned stiffness matrices. Taking advantages
from both strategies, the augmented Lagrangian method has become popular; see
e.g. Wriggers et al. [6] for an early presentation of the augmented Lagrangian

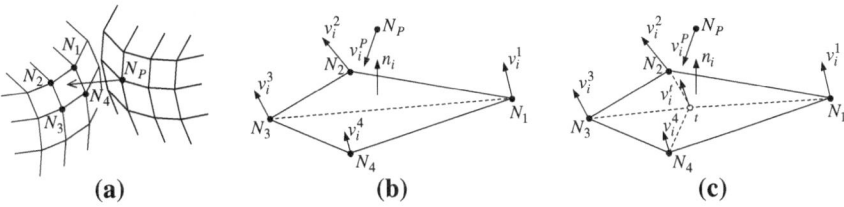

Fig. 4.3 Definition of contact pairs between deformable objects. **a** Node N_P contacting a quadrilateral element face $N_1 - N_2 - N_3 - N_4$ of another element. **b** Division of element face by diagonal. **c** Division of element face by temporary center node t

method. This method does however imply longer computation time than the pure penalty method due to iterations involving solution of the main system of equations in order to find the Lagrange multipliers. These iterations do not always converge fast, cf. Zavarise and Wriggers [7] who proposed an improved convergence scheme. Fast convergence is particularly critical for complex finite element computer programs involving non-linearities due to mechanical, thermal and electrical constitutive models. Many solutions assume frictionless or sticking contact, but friction has been included as well. Among the pioneers in frictional modeling are Simo and Laursen [8] using the augmented Lagrangian method.

In relation to resistance welding, Song et al. [9, 10] modeled contact in two dimensions by the penalty method. The contact between deformable objects in three dimensions to be presented in this section follows the work of Nielsen et al. [11] and is based on penalties for avoiding penetration of one object into another object or self-penetration of an object. All boundary nodes are analyzed for potential contact to another element face in each simulation step. If a certain node and a corresponding element face are identified as a potential contact pair, a normal gap velocity g_n^c is set up, such that if it is positive, the given velocity field will result in a gap in the contact pair, and if it is negative, the velocity field will result in penetration of the node and the element face. Depending on the mesh and the contact conditions, a node may be a contacting node in one contact pair, and at the same time take part in target surfaces in other contact pairs. This introduces symmetry in the contact algorithm naturally.

Figure 4.3a shows an example of a node N_P contacting an element face $N_1 - N_2 - N_3 - N_4$ of another element, in this case from another object. Identification of such contact pairs is based on a distance criterion by a small tolerance and that the relative velocity of N_P to the element face is orthogonally projecting to the element face. The definition of a plane is necessary from the element face in order to evaluate the orthogonal projection, but from four nodes, it generally does not exist, since a plane is defined by only three points. Therefore, the quadrilateral surface element is divided into triangles by one of the following two algorithms:

- *Algorithm I*: The face is divided into two triangles by division through a diagonal as shown in Fig. 4.3b. Doghri et al. [12] experienced loss of symmetry when applying this method. Division by a diagonal leaves two choices, thus resulting in two potential pairs of two triangles, and the problem is which pair to choose. In the present work, both divisions are evaluated, resulting in two overlapping triangles both containing the contact node projection. Among these, the triangle where the projection point results in most equal area coordinates is chosen. Area coordinates are defined as

$$\alpha_j = \frac{A_j}{\sum_{i=1}^{3} A_i} \tag{4.2.1}$$

with areas A_i defined in Fig. 4.1b. This selection of triangle has resulted in better representation of symmetry.

- *Algorithm II*: The face is divided into four triangles by a temporary center node t in the face as shown in Fig. 4.3c. This method was adopted by Doghri et al. [12] to overcome their loss of symmetry with the above method due to the unique choice of triangle. This algorithm is computationally more demanding due to larger expansion of the skyline of the stiffness matrix as the target face is represented by all four nodes compared to three nodes in the above algorithm.

When applying algorithm I, the normal gap velocity for contact pair c is defined as

$$g_n^{c(I)} = \left(v_i^P - \alpha_j v_i^j \right) n_i \tag{4.2.2}$$

where α_j are the area coordinates (4.2.1), v_i^j is the velocity of the j'th node of the selected triangle, and n_i is the normal to the triangle spanned by three of the element face nodes. Note the summation in i and j. Similarly for algorithm II, the normal gap velocity becomes

$$g_n^{c(II)} = \left(v_i^P - \alpha_1 v_i^1 - \alpha_2 v_i^2 - \alpha_t v_i^t \right) n_i \tag{4.2.3}$$

where index t refers to the temporary center node. Approximation of the velocity in the temporary center node by linear interpolation from the four face nodes, i.e. averaging, leads to

$$g_n^{c(II)} = \left(v_i^P - \left(\alpha_1 + \frac{\alpha_t}{4} \right) v_i^1 - \left(\alpha_2 + \frac{\alpha_t}{4} \right) v_i^2 - \frac{\alpha_t}{4} v_i^3 - \frac{\alpha_t}{4} v_i^4 \right) n_i \tag{4.2.4}$$

The normal gap velocity can be written in compact notation to ease subsequent derivations of the variational contribution to the energy rate functional. The following parameters are introduced for algorithms I and II to assist the compact notation,

$$\alpha_m^I = \begin{cases} 1 & \text{for } m = P \\ -\alpha_m & \text{for } m = 1, 2, 3 \end{cases} \tag{4.2.5}$$

$$A_I = \text{diag} \left\{ \alpha_P^I, \alpha_P^I, \alpha_P^I, \alpha_1^I, \alpha_1^I, \alpha_1^I, \alpha_2^I, \alpha_2^I, \alpha_2^I, \alpha_3^I, \alpha_3^I, \alpha_3^I \right\} \tag{4.2.6}$$

$$\mathbf{v}_I^T = \{\mathbf{v}_P, \mathbf{v}_1, \mathbf{v}_2, \mathbf{v}_3\}^T \tag{4.2.7}$$

$$\mathbf{n}_I^T = \{\mathbf{n}, \mathbf{n}, \mathbf{n}, \mathbf{n}\}^T \tag{4.2.8}$$

$$\alpha_m^{II} = \begin{cases} 1 & \text{for } m = P \\ -\left(\alpha_m + \frac{\alpha_t}{4}\right) & \text{for } m = 1, 2 \\ -\frac{\alpha_t}{4} & \text{for } m = 3, 4 \end{cases} \tag{4.2.9}$$

$$A_{II} = \text{diag} \left\{ \alpha_P^{II}, \alpha_P^{II}, \alpha_P^{II}, \alpha_1^{II}, \alpha_1^{II}, \alpha_1^{II}, \alpha_2^{II}, \alpha_2^{II}, \alpha_2^{II}, \alpha_3^{II}, \alpha_3^{II}, \alpha_3^{II}, \alpha_4^{II}, \alpha_4^{II}, \alpha_4^{II} \right\} \tag{4.2.10}$$

$$\mathbf{v}_{II}^T = \{\mathbf{v}_P, \mathbf{v}_1, \mathbf{v}_2, \mathbf{v}_3, \mathbf{v}_4\}^T \tag{4.2.11}$$

$$\mathbf{n}_{II}^T = \{\mathbf{n}, \mathbf{n}, \mathbf{n}, \mathbf{n}, \mathbf{n}\}^T \tag{4.2.12}$$

The velocity gap functions can then be written in the following compact notation for each of the algorithms ϕ,

$$g_n^{c(\phi)} = \mathbf{v}_\phi^T A_\phi \mathbf{n}_\phi \quad \phi = I, II \tag{4.2.13}$$

which by definition are equivalent to (4.2.2) and (4.2.4).

4.2.1 Frictionless Contact

According to the definition, action has to be taken only when $g_n^c < 0$ corresponding to penetration in the contact pair. In these cases, the velocity field is constrained by penalizing the penetration, through

$$\delta \Pi_C = \sum_{c=1}^{N_c} P g_n^c \delta g_n^c \tag{4.2.1.1}$$

which is to be added to the variation of the energy rate functional (3.2.1.11). The total number of contact pairs to be constrained is N_c, and P is a large positive constant. Equation (4.2.1.1) handles frictionless contact. In order to handle friction or full sticking, tangential velocity terms should be included.

Evaluation of (4.2.1.1) is accomplished for both algorithms by inserting the gap velocity g_n^c, while at the same time replacing $\mathbf{v}$ by $\mathbf{v}_0 + \Delta\mathbf{v}$ resembling the incremental finite element solution when using Newton–Raphson iterations. It is also noted that $g_n^{c(\varphi)} = \mathbf{n}_\varphi^T A_\varphi \mathbf{v}_\varphi$ is equivalent to (4.2.13), since it is simply the transpose of a scalar. Note also that $A_\varphi^T = A_\varphi$, since A_φ is a diagonal matrix. The substitution is shown in the following, where it has been utilized that the variation of the constant $\mathbf{v}_0$ is zero and $g_n^c \delta g_n^c = \delta g_n^c g_n^c$,

$$\delta \Pi_C = \sum_{c=1}^{N_c} P \delta \Delta \mathbf{v}_\varphi^T A_\varphi \mathbf{n}_\varphi \mathbf{n}_\varphi^T A_\varphi \left(\mathbf{v}_{0\varphi} + \Delta\mathbf{v}_\varphi \right), \quad \varphi = I, II \tag{4.2.1.2}$$

Fig. 4.4 Position of penalty terms in global system of equations with random order of the numbers P, 1, 2, 3, (4) as the position depends on their relative node numbers. All positions related to 4 are in parentheses as they are only active for contact algorithm $\phi = II$

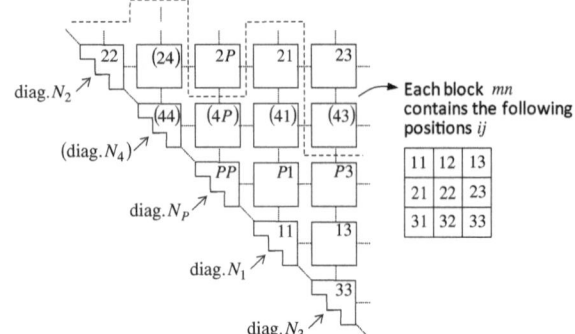

Utilizing that $\delta \Delta \mathbf{v}_\phi$ is to be chosen arbitrarily, it is possible to recognize the contributions to the stiffness matrix and the load vector after rearranging terms,

$$\delta \Pi_C = \sum_{c=1}^{N_c} \left[\delta \Delta \mathbf{v}_\varphi^T \left(\underbrace{P\mathbf{A}_\varphi \mathbf{n}_\varphi \mathbf{n}_\varphi^T \mathbf{A}_\varphi}_{\equiv \mathbf{K}_c} \Delta \mathbf{v}_\varphi + \underbrace{P\mathbf{A}_\varphi \mathbf{n}_\varphi \mathbf{n}_\varphi^T \mathbf{A}_\varphi \mathbf{v}_{0\varphi}}_{\equiv -\mathbf{f}_c} \right) \right], \quad \varphi = I, II$$

$$(4.2.1.3)$$

The contribution from the c'th contact pair to the stiffness matrix is $\mathbf{K}_c$, and the corresponding contribution to the load vector is $\mathbf{f}_c$. For algorithm I, the dimensions will be 12×12 for $\mathbf{K}_c$ and 12×1 for $\mathbf{f}_c$, whereas for algorithm II, the dimensions will be 15×15 and 15×1, respectively.

Regarding the assembly, an overview is best given by writing the contributions to the stiffness matrix and the load vector in the following forms, where for the load vector it is recognized that the initial gap velocity is $g_{n0}^c = \mathbf{n}_\varphi^T \mathbf{A}_\varphi \mathbf{v}_{0\varphi}$,

$$K_c^{ijmn} = P\alpha_m^\varphi \alpha_n^\varphi n_i n_j, \quad i, j = 1, 2, 3$$

$$m, n = \begin{cases} P, 1, 2, 3 & \text{for } \varphi = I \\ P, 1, 2, 3, 4 & \text{for } \varphi = II \end{cases} \qquad (4.2.1.4)$$

$$f_c^{jm} = \begin{cases} -P\alpha_m^\varphi n_j g_{n0}^c, & \text{for Newton–Raphson iterations} \\ 0, & \text{for direct iterations} \end{cases}$$

$$(4.2.1.5)$$

$$m, n = \begin{cases} P, 1, 2, 3 & \text{for } \varphi = I \\ P, 1, 2, 3, 4 & \text{for } \varphi = II \end{cases}$$

When using direct iterations, $\mathbf{v}$ is solved directly, rather than the incremental velocity $\Delta \mathbf{v}$ when using Newton–Raphson iterations. For direct iterations it follows (similar to (4.2.1.2) and (4.2.1.3)) that $\mathbf{K}_c$ is identical, but $\mathbf{f}_c = \mathbf{0}$.

The factor P is the penalty, α_m^φ is given by either (4.2.5) or (4.2.9), and n_i is the unit normal vector to the contact face. The position of each of the components, K_c^{ijmn} and f_c^{jm}, in the global system of equations is shown by Fig. 4.4, where blocks of 3×3 positions, ij, are identified as the relation between nodal points m and n. Each block is symmetric ($ij = ji$) and the blocks are symmetric around the main diagonal ($mn = nm$). In cases ($mn = \{2P, 23, (43)\}$ in Fig. 4.4) where the penalty blocks lie above the skyline, the skyline profile has to be expanded to allow the additional penalty blocks.

4.2.2 Sticking Contact

In sticking contact there is no sliding between the surfaces in contact. The tangential velocity difference is therefore penalized in addition to the normal gap velocity. The variational penalty term stemming from the tangential velocity difference g_t^c is given by

$$\delta \Pi_C^t = \sum_{c=1}^{N_c} P g_t^c \delta g_t^c \tag{4.2.2.1}$$

which is similar to (4.2.1.1). The derivations are also identical except that the tangential contributions result in two sets of penalty terms corresponding to the two tangential components of the tangential velocity difference written as

$$g_{t1}^{c(\varphi)} = \mathbf{v}_\varphi^T \mathbf{A}_\varphi \mathbf{t}_{\varphi 1}, \quad g_{t2}^{c(\varphi)} = \mathbf{v}_\varphi^T \mathbf{A}_\varphi \mathbf{t}_{\varphi 2}, \quad \varphi = I, II \tag{4.2.2.2}$$

with notations following (4.2.5)–(4.2.12) and $\mathbf{t}_{\varphi 1}$ and $\mathbf{t}_{\varphi 2}$ being vectors of the two tangential unit vectors. The resulting terms after insertion into (4.2.2.1) are similar to (4.2.1.4)–(4.2.1.5) with the normal vector exchanged by each of the tangential vectors.

4.2.3 Frictional Contact

As for the contact between workpiece and rigid tools, the constant friction law, $\tau_f = mk$, will be taken as an example. To avoid the derivatives going to infinity cf. the discussion related to Fig. 4.2, the friction stress is written as (4.1.4), here with the tangential velocity difference defined by (4.2.2.2),

$$\tau_f \cong -\frac{2}{\pi} mk \cdot \arctan \left(\frac{g_t^{c(\varphi)}}{u_0} \right) \tag{4.2.3.1}$$

The contribution to the energy rate functional due to friction and its corresponding variation are

$$\Pi_f = \sum_{c=1}^{N_c} \tau_f A_c g_t^c \tag{4.2.3.2}$$

$$\delta \Pi_f = \sum_{c=1}^{N_c} \left[\delta \tau_f A_c g_t^c + \tau_f A_c \delta g_t^c \right] \tag{4.2.3.3}$$

where frictional force is introduced by the product of the frictional stress τ_f and the area of the contact pair A_c. Insertion of the friction stress and the tangential velocity difference components (4.2.2.2) into the variational form (4.2.3.3) results in the following additional terms to the stiffness matrix K_f^{ijmn} and generalized load vector f_f^{jm} for each of the tangential components (one set for each of the inserted tangential unit vectors),

$$K_f^{ijmn} = -\frac{2}{\pi} mk \frac{u_0}{(g_t^c)^2 + u_0^2} \alpha_m^\varphi \alpha_n^\varphi t_i t_j, \quad i,j = 1,2,3$$

$$m,n = \begin{cases} P,1,2,3 & \text{for } \varphi = I \\ P,1,2,3,4 & \text{for } \varphi = II \end{cases} \tag{4.2.3.4}$$

$$f_f^{jm} = \begin{cases} -\frac{2}{\pi} mk A_c \left(\frac{u_0 g_{t0}^c}{(g_t^c)^2 + u_0^2} + \arctan\left(\frac{g_t^c}{u_0}\right) \right) \alpha_m^\varphi t_j, & \text{for Newton–Raphson iterations} \\ -\frac{2}{\pi} mk A_c \arctan\left(\frac{g_t^c}{u_0}\right) \alpha_m^\varphi t_j, & \text{for direct iterations} \end{cases}$$

$$m = \begin{cases} P,1,2,3 & \text{for } \varphi = I \\ P,1,2,3,4 & \text{for } \varphi = II \end{cases} \tag{4.2.3.5}$$

4.2.4 Electrical and Thermal Contact

Electrical and thermal contact properties are included in contact interface elements on one or both of the objects in contact and eventual drops over the interface due to contact resistances are included in these elements, see Chap. 7 for a description of the physical properties. The contact implementation here is therefore limited to ensure that the electrical potential and the temperature are identical on both sides of the contacting finite elements. This is ensured by penalizing electrical potential difference Φ_d and temperature difference T_d by

$$\delta \Pi_\Phi = \sum_{c=1}^{N_c} P \Phi_d^c \delta \Phi_d^c \quad \delta \Pi_T = \sum_{c=1}^{N_c} P T_d^c \delta T_d^c \tag{4.2.4.1}$$

Both the potential and the temperature are scalar fields, and the derivation is therefore a reduced form of the frictionless contact derivation in the absence of the normal vector. The contributions to the system matrices are

$$K_{\Phi,T}^{mn} = P \alpha_m^\phi \alpha_n^\phi, \quad m,n = \begin{cases} P,1,2,3 & \text{for } \phi = I \\ P,1,2,3,4 & \text{for } \phi = II \end{cases} \tag{4.2.4.2}$$

while there are no contribution to the right hand sides.

4.3 Application of Contact Modeling

The numerical simulation of the forming process utilized for producing small size, seamless, cylindrical metallic reservoirs requires successful utilization of both contact modeling approaches that were presented in previous sections of this chapter. The forming process is schematically shown in Fig. 4.5 and consists of: (i) upper and lower semi-ellipsoidal shaped dies, (ii) a container, (iii) a mandrel and (iv) a tubular preform [13, 14].

The raw materials utilized in the fabrication of the reservoirs consisted of commercial tubes of aluminum AA6063-T0 and internal mandrels made from a commercial low melting point alloy MCP137 ($T_{melt} = 137\,°C$) comprising bismuth, lead, tin and cadmium.

As seen in Fig. 4.5, the forming operation is accomplished by axial pressing of the open ends of a tubular preform with the upper semi-ellipsoidal shaped die until achieving the desired geometry. The upper and lower dies are the active tool components and its sharp-edges are protected against collapse due to circumferential tensile stresses by means of the container which acts as a shrink fit tool part. The dies are dedicated to a specific outer radius of the tube r_0 and its profile defines the geometry of the reservoir. The container constrains material from outward flow in order to avoid the occurrence of buckling and helps minimizing the errors due to

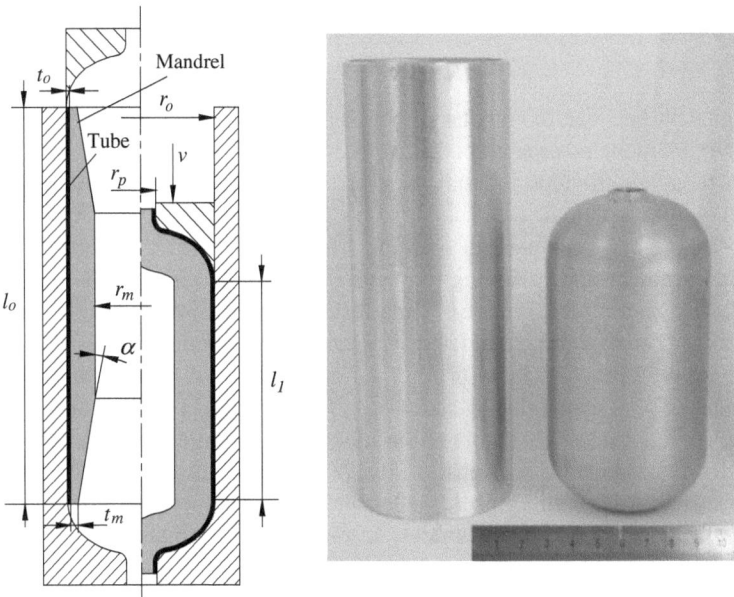

Fig. 4.5 Shaping a tubular preform into a small-size cylindrical reservoir with semi-ellipsoidal ends by cold forming. The enclosed photograph shows the preform and the final reservoir made from Aluminum AA6063-T0 with 60 mm diameter

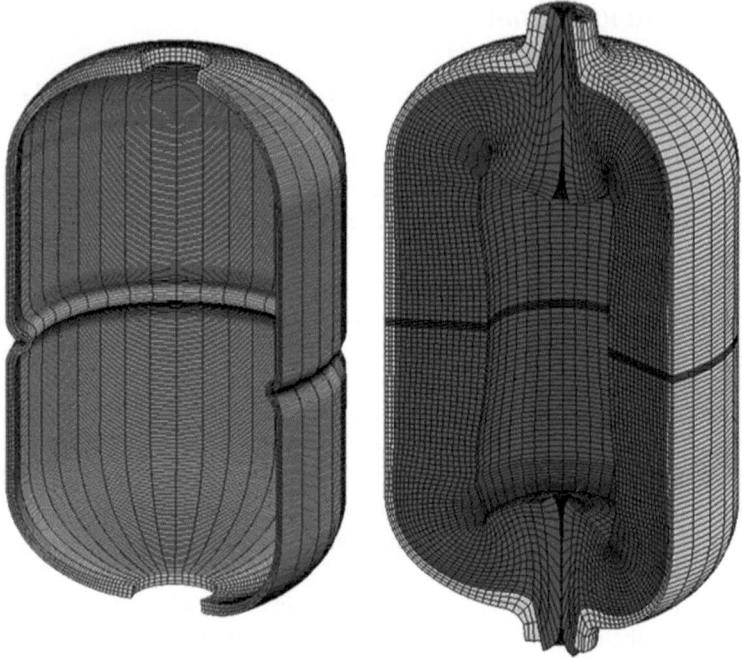

Fig. 4.6 Forming tubular preforms into cylindrical reservoirs with semi-ellipsoidal ends. Finite element predicted geometry at the end of the process without and with internal deformable mandrel

misalignment between the tubular preforms and the individual dies. Both dies and container are modeled as rigid bodies.

The mandrel provides internal support to the tubular preform during plastic deformation in order to avoid collapse by wrinkling and local instability at the equatorial region. The mandrel is made from a low melting point alloy that is capable of continuously adapting its shape to that of the formed tube and is easily removed by melting (recyclable), while leaving the reservoir intact, at the end of the process. The mandrel is modeled as a deformable body.

The forming process shown in Fig. 4.5 is the result of four basic mechanisms that compete with each other; plastic work, friction, local buckling and wrinkling. Plastic work is caused by compression along the circumferential direction which gradually deforms the tube against the dies. Friction develops gradually as the tube deforms against the semi-ellipsoidal shaped dies. Local buckling and wrinkling are associated with compressive instability in the axial and circumferential directions and limit the overall formability of the process by giving rise to non-admissible modes of deformation.

Figure 4.6 shows the computed predicted geometry of the reservoir at the end of the process. As can be seen, the interaction between the tube and the internal deformable mandrel allows fabricating sound reservoirs whereas forming without a mandrel will inevitably lead to failure. Subsequent removal of the internal mandrel by melting, while leaving the shell intact, results in the reservoir shown in Fig. 4.5. This example puts into evidence the critical role played by contact algorithms in ensuring adequate estimates of plastic flow.

References

1. Barata Marques MJM, Martins PAF (1990) Three-dimensional finite element contact algorithm for metal forming. Int J Numer Methods Eng 30:1341–1354
2. Chenot JL (1987) Finite element calculation of unilateral contact with friction in non-steady-state processes. In: Proceedings of the NUMETA Conference, Swansea, UK
3. Shiau YC, Kobayashi S (1988) Three dimensional finite element analysis of open die forging. Int J Numer Methods Eng 25:67–86
4. Yoon JH, Yang DY (1988) Rigid plastic finite element analysis of three dimensional forging by considering friction on continuous curved dies with initial guess generation. Int J Mech Sci 30:887–898
5. Chen C, Kobayashi S (1978) Rigid plastic finite element analysis of ring compression. Application of numerical methods in forming processes, AMD 28. ASME, New York, pp 163–174
6. Wriggers P, Simo JC, Taylor RL (1985) Penalty and augmented Lagrangian formulations for contact problems. In: Proceedings of the NUMETA Conference, pp 97–105
7. Zavarise G, Wriggers P (1999) A superlinear convergent augmented Lagrangian procedure for contact problems. Eng Comput 16:88–119
8. Simo JC, Laursen TA (1992) An augmented Lagrangian treatment of contact problems involving friction. Comput Struct 42:97–116
9. Song Q, Zhang W, Bay N (2006) Contact modelling in resistance welding. Part 1: algorithms and numerical verification. J Eng Manuf 220(5):599–606
10. Song Q, Zhang W, Bay N (2006) Contact modelling in resistance welding. Part 2: experimental validation. J Eng Manuf 220(5):607–613
11. Nielsen CV, Martins PAF, Zhang W, Bay N (2011) Mechanical contact experiments and simulations. Steel Res Int 82:645–650
12. Doghri I, Muller A, Taylor RL (1998) A general three-dimensional contact procedure for implicit finite element codes. Eng Comput 15:233–259
13. Alves LM, Martins PAF, Pardal TC, Almeida PJ, Valverde NM (2009) Plastic deformation technological process for production of thin-wall revolution shells from tubular billets. Patent request no. PCT/PT2009/000007, European Patent Office
14. Alves LM, Silva MB, Martins PAF (2011) Fabrication of small size seamless reservoirs by tube forming. Int J Press Vessels Pip 88:239–247

Chapter 5
Meshing and Remeshing

A significant amount of time in finite element modeling of manufacturing processes is spent in mesh generation. Setting up three-dimensional meshes is a cumbersome task due to complexity of the processes and the involved geometries. Moreover, additional meshing challenges often appear due to the fact that manufacturing processes based on large plastic deformations present progressive mesh distortion (or degeneracy), potential interference between mesh and contour of the tools and possible contact of the mesh with itself. This poses the need for robust, automatic, mesh generation and regeneration (remeshing) procedures in order to ensure that complex processes are modeled from the beginning to the end with high levels of accuracy both in terms of geometry and distribution of field variables.

The choice of element type has large impact on the simulations, and the typical dilemma in three dimensions arises from the selection between tetrahedral and hexahedral elements. The arguments for the tetrahedral elements are the robustness, versatility and availability of meshing algorithms. Based on Delaunay tessellation, Coupez et al. [1] opened the possibility of effectively and automatically simulating the whole forming process of complex three-dimensional parts from beginning to the end. On the other hand, the argument for the hexahedral elements is the accuracy. Furthermore, standard tetrahedral elements suffer from locking due to the incompressibility constraint in plasticity. Second-order tetrahedral elements overcome this problem but perform poorly in the tool-workpiece contact interfaces, often leading to stability problems in the contact algorithms as stated by Tekkaya and Martins [2]. As a result of this, special tetrahedral elements with interior nodes have been developed for preventing locking. These elements, however, still suffer from some of the typical drawbacks of tetrahedral elements: They are overly stiff, very sensitive to mesh orientation and frequently require up to an order of magnitude more elements to achieve the same level of accuracy as hexahedral elements. Benzley et al. [3] also noticed that meshes based on tetrahedral elements result in larger models, and therefore in larger computational requirements, than meshes based on hexahedra for the same level of accuracy. Kraft [4] observed that tetrahedral elements cause critical errors when distorted,

C. V. Nielsen et al., *Modeling of Thermo-Electro-Mechanical Manufacturing Processes*, SpringerBriefs in Applied Sciences and Technology, DOI: 10.1007/978-1-4471-4643-8_5, © The Author(s) 2013

whereas hexahedra have better behavior even when distorted. This chapter deals with hexahedral elements due to the challenges currently posed by all-hexahedral meshing and remeshing algorithms and computer implementation.

Meshes based on hexahedral elements can be divided into two groups. One group is structured meshes, which can be recognized by all interior nodes of the mesh having equal number of adjacent elements. The simplest geometries are easily meshed and the more complicated can be handled by isoparametric meshing of superelements as described in Sect. 5.2. The utilization of this method is limited to geometries that can be divided into hexahedral superelements. The second group is unstructured meshes, which, in principle, should cover all three-dimensional geometries.

The simplest unstructured meshing by hexahedral elements is performed by means of an indirect approach, where the geometry is first meshed by tetrahedral elements using Delaunay tessellation. Each tetrahedron is subsequently decomposed into four hexahedral elements. This approach is robust but always leads to distorted elements with only a fraction of the quality of an ideal hexahedron. Furthermore, the indirect meshing by decomposition always leads to nodal points with high valence, which artificially increases the overall stiffness of the finite element models. The poor quality obtained by this approach is considered the reason why some well-known commercial finite element programs currently utilized in metal forming do not offer hexahedral elements as an option, or do not provide automatic remeshing if hexahedra are available. The alternative approach for the automatic generation of good quality hexahedral elements in arbitrary domains was originally proposed by Schneiders and Bünten [5] and it will be hereafter named as "all-hexahedral meshing."

All-hexahedral meshing is presented in Sect. 5.3 and all-hexahedral remeshing is presented in Sect. 5.4. Because description of tooling is relevant for meshing and remeshing, a brief review of the techniques that are utilized for the description of tool surfaces is given in Sect. 5.1.

5.1 Description of Tooling

Tools can be described by analytical or parametric surfaces, surface meshes and clouds of points, Santos and Makinouchi [6]. In most of the commercial finite element computer programs, the surfaces are described by means of surface meshes (e.g. triangular elements, Fig. 5.1a). The utilization of a grid of triangular elements instead of alternative approaches based on analytical functions, parametric surfaces (Fig. 5.1b) or clouds of points, is due to the fact that the former always guarantees successful discretization of the surfaces while other techniques often face difficulties whenever complex shapes and/or small geometrical details are to be discretized.

However, triangular elements fail to ensure smoothness and, therefore, introduce artificial roughness on the surface of tooling. This can bring in geometrical

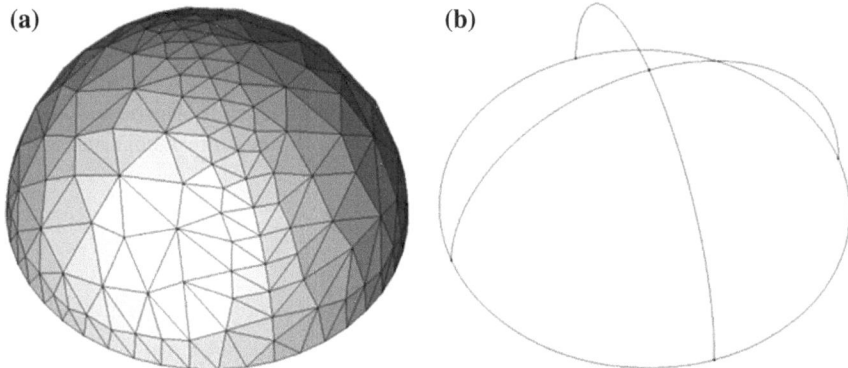

Fig. 5.1 Two main approaches utilized in the discretization of a hemispherical tool: **a** surface meshes and **b** analytical functions or parametric surfaces

errors, for instance in case of small fillet radii which may be poorly captured if the discretization is too coarse.

Tools deform elastically but are commonly modeled as rigid in three-dimensional finite element modeling of metalworking processes because their deformations are negligible when compared to the plastic deformation of the workpieces. However, the simplification is not always feasible as shown by Tekkaya and Martins [2] by means of a metal forming example displaying significant erroneous tool force when assuming the tool rigid instead of elastic. In such situations there is a need to take the elastic deformation of tooling into consideration. When it comes to resistance welding applications, the tools are acting as the coupling between electrodes and the welding machine and are therefore sufficiently modeled as rigid.

The majority of the applications reported in the literature that deal with the elastic deformation of tools is restricted to the utilization of finite elements both in the workpiece material and tools, Boussetta et al. [7] and Behrens and Kerkeling [8]. This results in limitations in terms of the size and complexity of the overall computer models when the tools, having complex geometrical shapes, are to be discretized and included in the overall set of finite-element computations. Some of these limitations can be solved by alternative approaches based on combination of finite element and boundary element methods; see Fernandes et al. [9].

The utilization of boundary elements for performing the elastic deformation of the dies not only avoids over-sizing the resulting computer models as it offers significant computational advantages over the existing approaches fully based on finite elements. The first advantage is due to the fact that boundary elements only require discretization of the die surfaces. The second advantage is seen by taking into consideration that numerical simulation of manufacturing processes is generally accomplished through a succession of displacement increments, each modeling a small percentage of the initial height of the preform. In practical terms, this means that a simulation based on several hundreds of increments

will require the elastic deformation of the dies to be also calculated hundreds
of times. This is the reason why alternative approaches based on boundary elements make a difference against fully finite element based solutions. Similar
finite element—boundary element combined approaches can also be utilized for
solving thermo-mechanical coupling in the tool-workpiece interface, as shown by
Rodrigues and Martins [10].

5.2 Isoparametric Structured Meshing

Structured meshes of hexahedral elements can be created by a method based
on isoparametric meshing of superelements as first shown by Zienkiewicz and
Phillips [11]. Martins and Barata Marques [12] developed a three-dimensional
mesh generator based on this technique and published the source code.

The method is applicable when the geometry to be meshed can be divided into
a number of sub-blocks, the so-called superelements. An example is shown in
Fig. 5.2a in terms of a quarter of an electrode for spot welding. The top face shows
a typical division of solid cylindrical faces in order to achieve well shaped superelements resulting in well-shaped 8-node hexahedral elements.

The superelements are 20-node elements specified by the user by the coordinates of the eight corner points and 12 mid-side points. The mid-side points are
automatically placed half distance on the straight line between two corner points if
not specified. Otherwise, the edges of the superelements are represented parabolic
by the mid-side nodes and their two respective corner points. Any point within the
superelements is given by interpolation using the standard shape functions for a

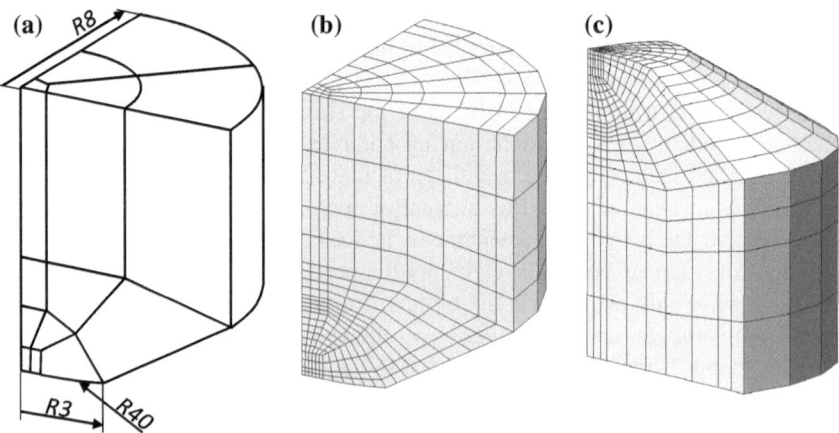

Fig. 5.2 Isoparametric structured meshing of a quarter of an ISO type B0 electrode for spot
welding. **a** Subdivision of geometry into 20-node superelements. **b, c** Subsequent division of
superelements into 8-node hexahedral elements

20-node hexahedral element. Division of the superelements into 8-node elements is based on specified number of divisions along each superelement side along with a corresponding grading of the element division. Note that divisions should only be specified for the three mutual orthogonal (with respect to natural coordinates) directions due to the structured nature of the created meshes. Figures 5.2b, c show an example of the resulting mesh.

In combination with a user interface for setting up the superelements, this meshing technique is a powerful tool in setting up initial geometries and meshes. The method is not automatically applicable in remeshing procedures because the underlying geometries of the meshes with need for remeshing usually cannot be identified by a reasonable number of superelements. Remeshing is therefore solely accomplished by meshing techniques based on unstructured meshing.

5.3 All-Hexahedral Unstructured Meshing

All-hexahedral meshing is a grid based approach that involves the construction of a structured three-dimensional mesh of hexahedra in the interior of the volume (core mesh) followed by subsequent generation of an extra layer of elements for linking the core with its projection on the boundary of the workpiece. The method proposed by Schneiders and Bünten [5] is an extension of the two-dimensional approach based on quadrilateral elements that was previously developed by Schneiders et al. [13]. Among other contributors to the all-hexahedral meshing techniques are e.g. Kraft [4], Zhu and Gotoh [14], Karadogan and Tekkaya [15] and Kwak and Im [16].

The all-hexahedral meshing algorithm to be presented in what follows was originally developed by Fernandes and Martins [17], who provided a detailed description of the major procedures and programming solutions, and further developed by Nielsen et al. [18], who included adaptive core meshes and the possibility of handling multiple objects besides enhancing the overall robustness and versatility.

5.3.1 Identification of Geometric Features and Selection of Core Mesh

The starting point is a triangular surface mesh of the geometry, e.g. provided by a CAD program, and the meshing procedure is then responsible for supplying a hexahedral mesh within the surface. An important step before the meshing itself is the recognition of geometrical features in form of vertices and edges that must be kept during meshing. Figure 5.3a shows a triangular surface mesh and Fig. 5.3b shows the geometrical features that were identified after applying algebraic algorithms

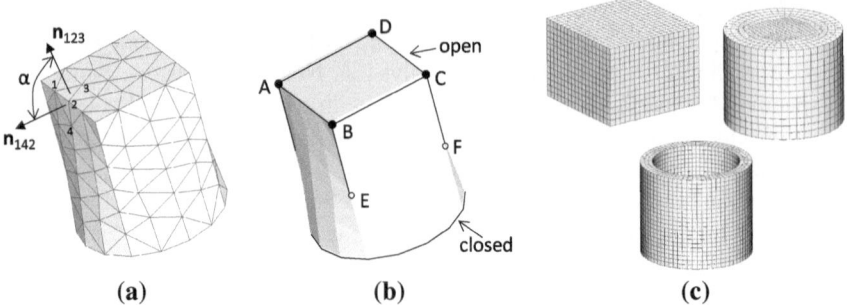

Fig. 5.3 Identification of geometrical features and examples of bounding boxes. **a** Identification of edge segments from triangular surface mesh. **b** Identification of edges and vertices. **c** Examples of bounding boxes

based on the evaluation of surface normals to the triangles and analyzing nodal valences. As illustrated in Fig. 5.3a, a typical segment $1 - 2$ shared by two adjacent triangular elements '123' and '142' is taken as an edge segment if the angle α between the normals $\mathbf{n}_{123}$ and $\mathbf{n}_{142}$ to the triangular elements is greater than a specified threshold angle (say $\theta = 45°$),

$$\mathbf{n}_{123} \cdot \mathbf{n}_{142} > cos\theta \qquad (5.3.1.1)$$

The summation and sorting of adjacent edge segments before and after $1 - 2$ in a sequential manner leads to the edge $A - B$ (Fig. 5.3b).

Vertices are collected from the end points of edge segments that are connected to at least three neighboring edge segments. Edges are classified into three main groups (Fig. 5.3b): (i) open edges (ii) closed edges and (iii) fading edges. Open edges connect two different vertices ($A - B$, $B - C$, $C - D$ and $D - A$), closed edges start and end in the same point and do not contain vertices, and fading edges start in a vertex but smoothly vanish along the surface (e.g. $B - E$ and $C - F$).

The first step of the meshing procedure is the generation of the core mesh of hexahedral elements. This is generated by creating a bounding box of elements as those shown in Fig. 5.3c and subsequently removing the elements with at least one node outside the provided surface. Selection of the bounding box depending on the dominant geometric primitive identified from the surface mesh was proposed by Nielsen et al. [18] and proved to increase mesh quality significantly. It is noted that the choice of bounding box is not limited to those shown in the figure, but could potentially be any mesh easily generated by means of isoparametric based procedures outlined in Sect. 5.2. The removal of elements outside the provided surface is accommodated by a ray-tracing algorithm described by O'Rourke [19] to judge if a node is inside or outside the surface. Following a vector in an arbitrary direction from a certain node the number of intersections with surface triangles determines if the node is inside (odd number of intersections) or outside (even number of intersections).

5.3.2 Reconstruction of Geometry

Reconstruction of the geometry includes introduction of an additional layer of elements on the core mesh with projection to the surface, projection of selected nodes to vertices and projection of nodes to reconstruct edges. The reconstruction of surfaces is successfully performed with the isomorphism technique originally proposed by Schneiders and Bünten [5] and later modified by Fernandes and Martins [17]. The isomorphism technique is based on the generation of a layer of elements between the core mesh and the triangular surface mesh that defines the contour of the workpiece. The core mesh is smoothened before projecting the outmost nodal points to the triangular surface mesh in order to avoid crossing of adjacent surface normals that would create projection problems.

Reconstruction of vertices is performed by projection of nodal points of the aforementioned layer of hexahedral elements to the vertices of the triangular surface mesh. The algorithm implemented by Nielsen et al. [18] is based on a combination of the usual distance criterion and the valence (the number of element edges attached to a node) in order to avoid creating degenerated elements that will need subsequent repairment. Moreover, the algorithm is built upon an iterative search for the best candidate to be projected to the vertex. The iterative searching procedure is important for reconstructing sharp corners, where the distance to the core mesh can be large and no candidates are likely found at first. On the other hand, the iterative procedure allows the search radius to be progressively increased from small values in order to avoid candidates located far away to be projected onto the existing vertex.

Edge reconstruction is the most critical step in all-hexahedral based meshing. The procedure is illustrated in Fig. 5.4 and is based on the algorithm by Kwak and Im [16] modified by Nielsen et al. [18] to include additional geometrical features and topology based constraints. Figure 5.4a shows the final mesh of the example and Fig. 5.4c–e show magnified details of the intermediate meshes. The mesh included in Fig. 5.4c was plotted after vertex reconstruction while the meshes in Fig. 5.4d, e were taken after partial (from vertex node V to edge node P) and final reconstruction of edges. It is important to notice that 'final reconstruction' of an edge should not be confused with its 'completeness', as can be easily observed in Fig. 5.4e. Completion of edges is discussed in Sect. 5.3.3.

In the selection of the best candidate to project to the edge after P, the first step is the identification and sorting of up to three candidate nodes based on the distance to the last projected node P. Nodal point $k_{(1)}$ is the first potential candidate, but it is discarded because its angle with the previous part of the edge is larger than a critical threshold value (say 60°), implying that $k_{(1)}$ is not considered as a node ahead of P. The next nodal point to be evaluated is k_1, which fulfills all the necessary conditions and is accepted as candidate number 1. Node $k_{(2)}$ is discarded as second candidate because it is diagonally opposed to P (that is, line segment drawn from $k_{(2)}$ to P is a diagonal of the quadrilateral face). Node k_2 is selected as candidate 2, and during selection of candidate 3, $k_{(3)}$ is discarded for also being diagonally opposed to P, and k_3 is selected instead. The introduction of the

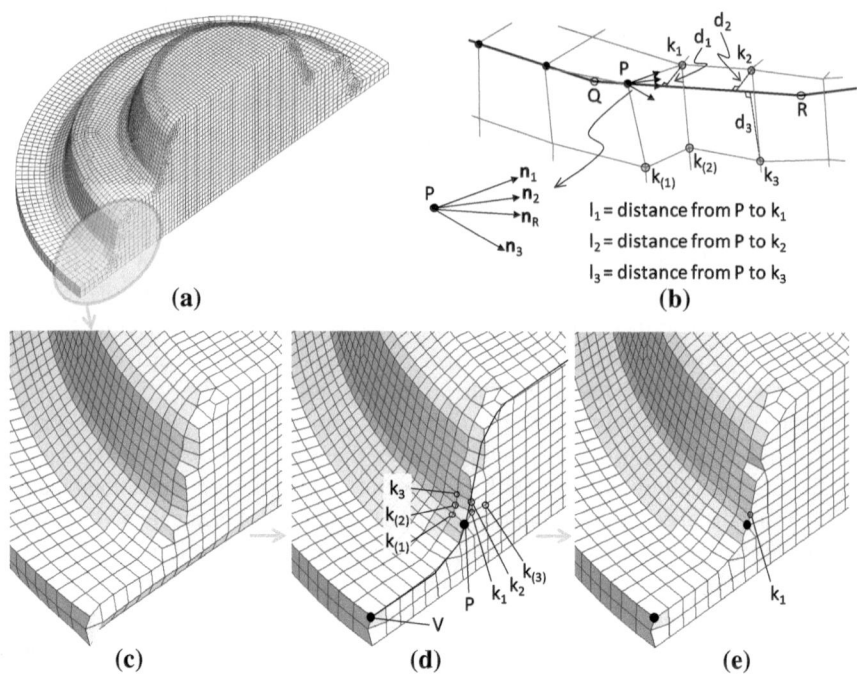

Fig. 5.4 Reconstruction of edges in a typical forged flange component. **a** Final hexahedral mesh. **b** Schematic illustration of reconstruction process on edge segment QR. **c** Mesh after reconstruction of vertices. **d** Mesh after partial reconstruction of edges. **e** Mesh after final reconstruction of edges showing k_1 projected on the edge and evidence of lack of completeness

topology based criterion avoiding diagonally opposite nodes on the edge prevents the occurrence of degenerated hexahedral elements along the edges. However, this type of constraint should not be confused with the necessity of having diagonals on the edges for ensuring its completeness as addressed in Sect. 5.3.3.

The second step is to choose between the three identified candidates. Figure 5.4b illustrates such a situation with respect to the latest projected node P for candidate nodes k_1, k_2 and k_3. The first priority is obtained after applying the following function (hereafter named g-function),

$$g = (1 - \mathbf{n}_R \cdot \mathbf{n}_i) + \frac{d_i}{\max{(d_1, d_2, d_3, C_0)}} + \frac{l_i}{\max{(d_1, d_2, d_3, C_0)}} \quad (5.3.2.1)$$

where $\mathbf{n}_R$ is the unit vector from P to R and $\mathbf{n}_i$ are the unit vectors from P towards the candidates. The distances from the candidates to P are denoted l_i and the distances from the candidates to the edge segment $Q - R$ of the triangular surface mesh are represented by d_i, where index i refers to the candidates. The constant C_0 refers to the characteristic element side of the core mesh. In choosing between candidates, first priority is given to the candidate minimizing the g-function while the candidate maximizing the g-function is directly discarded. In the example in

Fig. 5.4, candidate k_1 is selected from the minimization of the g-function, and the projection of k_1 is shown in Fig. 5.4e together with the remaining projections based on the application of the proposed algorithm.

The first two terms of the g-function were originally suggested by Kwak and Im [16] and account for the selection of candidates that minimizes collinearity and distance to the edge segment. The third term was added by Nielsen et al. [18] to force minimization of the g-function to be dependent on the distance to the latest projected node P. The importance of the new term is best illustrated by an example where two candidates are equidistant to a straight edge, such that the second term in (5.3.2.1) is of no importance. In this case the first term alone would prioritize the candidates further away from P due to collinearity, although the nearer node may fulfill all other criteria for being chosen. The third term adds robustness by compromising between collinearity and distance to node p.

5.3.3 Edge Repairment

As it was mentioned in relation to reconstruction of edges in Fig. 5.4 and exemplified further in Fig. 5.5 by an extreme geometry in form of a hexahedron, there is often necessity of performing repairment of the edges in order to ensure completeness of the edges. Figure 5.5a shows the core mesh generated from a cuboid bounding box and Fig. 5.5b shows the intermediate mesh after reconstruction of surfaces, vertices and edges.

Fig. 5.5 Selected overview of topology based repairment of edges in all-hexahedral meshing of a tetrahedron. **a** Core mesh obtained from a cuboid bounding box. **b** Mesh after reconstruction of surfaces, vertices and edges. **c** Mesh after topology based repairments illustrated by node pairs 1*a*, 1*b* and 2*a*, 2*b*. **d** Final mesh after application of templates for eliminating degenerated hexahedra (e.g. '3') by means of its decomposition into well-shaped hexahedra

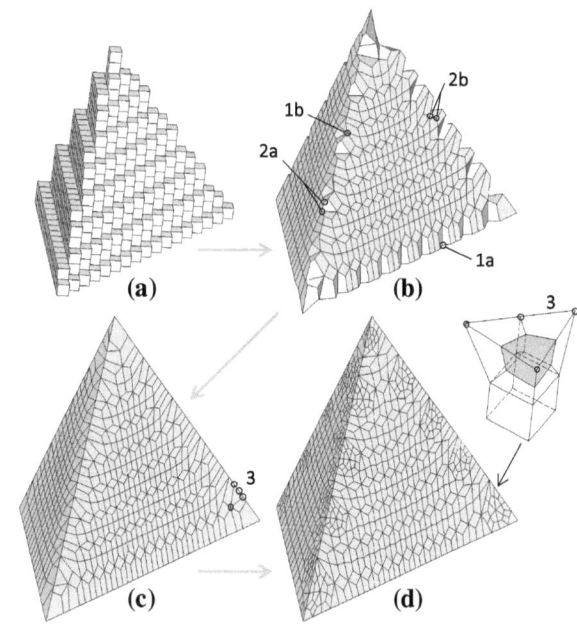

The necessity of repairment to complete the edges by element sides and to improve the element quality is obvious. Projection of nodes, such as 1*a* and 1*b* in Fig. 5.5b, is one of the topology based repairment procedures applied. Nodes 1*a* and 1*b* are characterized by being neighbors of two consecutive nodes on the edge that do not share an element side. Their projections will locally complete the corresponding edges. Another topology based repairment is the projection of neighboring pairs of nodes such as 2*a* and 2*b* in Fig. 5.5b. Each of these nodes is neighboring one of two consecutive nodes on the edge, which do not share an element side. Again, the repairment ensures local completion of the edge. The mesh of the tetrahedron after performing the two previously mentioned types of repairment is shown in Fig. 5.5c.

At this stage the edges are complete, but additional repairment is still necessary to resolve degenerated elements, such as that labeled '3' in the figure. The element has three nodes along the edge and can be split into four elements of better quality by means of the template proposed by Schneiders and Bünten [5]. The template is illustrated by the detail in Fig. 5.5d, where also the resulting hexahedral finite element mesh is shown.

5.3.4 Smoothing

Smoothing procedures are applied with the purpose of repairing distorted elements and improving their shape in different stages of meshing and remeshing. In general terms, smoothing is accomplished by changing the position of the nodal points to new positions given by a weighted average of the neighboring nodal points without modifying the topology of the mesh.

Several constraints must be taken into account to preserve the geometrical consistency of the hexahedral meshes. Vertices are excluded from smoothing as their positions are fixed. Edge nodes stay on the edges and the surface nodes remain on the surfaces. To overcome these constraints, edges are smoothed first by means of a parametric based procedure presented by Nielsen et al. [18]. Surfaces are smoothed next while excluding the edge nodes, and finally, the volume is smoothed while excluding all nodal points located on edges and surfaces. Surfaces are smoothed by averaging nodal positions according to the weighted areas of the neighboring surface quadrilaterals and volume smoothing is performed by averaging nodal positions of the core mesh according to weighted volumes of neighboring hexahedral elements. Both surface and volume smoothing procedures are comprehensively described elsewhere; see Karadogan and Tekkaya [15] and Fernandes and Martins [17].

5.3.5 Application of All-Hexahedral Meshing

A connecting rod is presented in Fig. 5.6 to illustrate the applicability of the presented all-hexahedral meshing technique. The shape of the connecting rod is provided by a triangularized surface from AutoCAD. Following the above

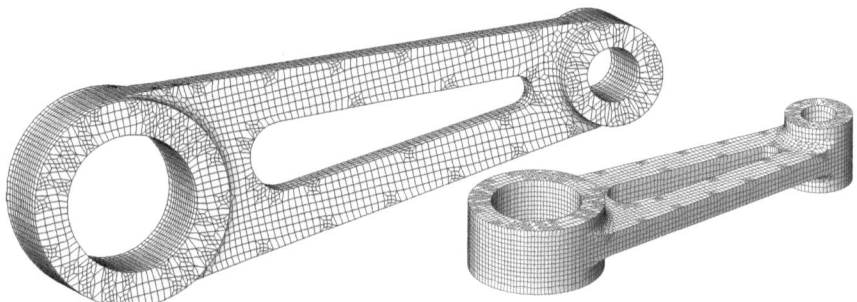

Fig. 5.6 Hexahedral mesh of a connecting rod showing the capabilities of the presented all-hexahedral meshing technique

presented procedures from identification of geometric features, generation of core mesh and reconstruction of edges, the smoothened mesh with improved element conditions is obtained as shown in Fig. 5.6.

5.4 Remeshing

The description of the remeshing procedures is based on the example shown in Fig. 5.7, which is an industrial case consisting of the resistance welding of a square nut to a sheet. Resistance welding is an extreme case of multi-object simulation involving electro-thermo-mechanical modeling as described in the present work, and in terms of remeshing it presents several complications due to multiple objects and local effects presented by the process.

Figure 5.7a shows the initial mesh of the case simulated by one quarter due to symmetries. A standard component in the automotive industry in form of an M10 steel square nut (1) is welded to an AISI 1008 steel sheet (2) of 1.4 mm thickness. A 50 μm thin layer of elements (3) on top of the sheet provides the interface properties between the square nut and the sheet. Electrical and thermal resistances stemming from oxide layers, surface films and contaminants are included in this layer. The electrical and thermal contact properties change with temperature and contact pressure due to formation of real contact area break down and squeeze out of the impurities. A 15 kA direct current is applied through the copper alloy electrodes (4) and (5) from each side of the square nut and the sheet.

The temperature field after simulating 80 ms of the resistance welding process is shown in the deformed mesh of Fig. 5.7b. Figure 5.7c, d show the temperature of the original and the remeshed cases after additionally 40 simulation steps, corresponding to 100 ms in total. Nearly identical shape and temperature distributions prove the accuracy, reliability and validity of the overall procedure, which is outlined in the following.

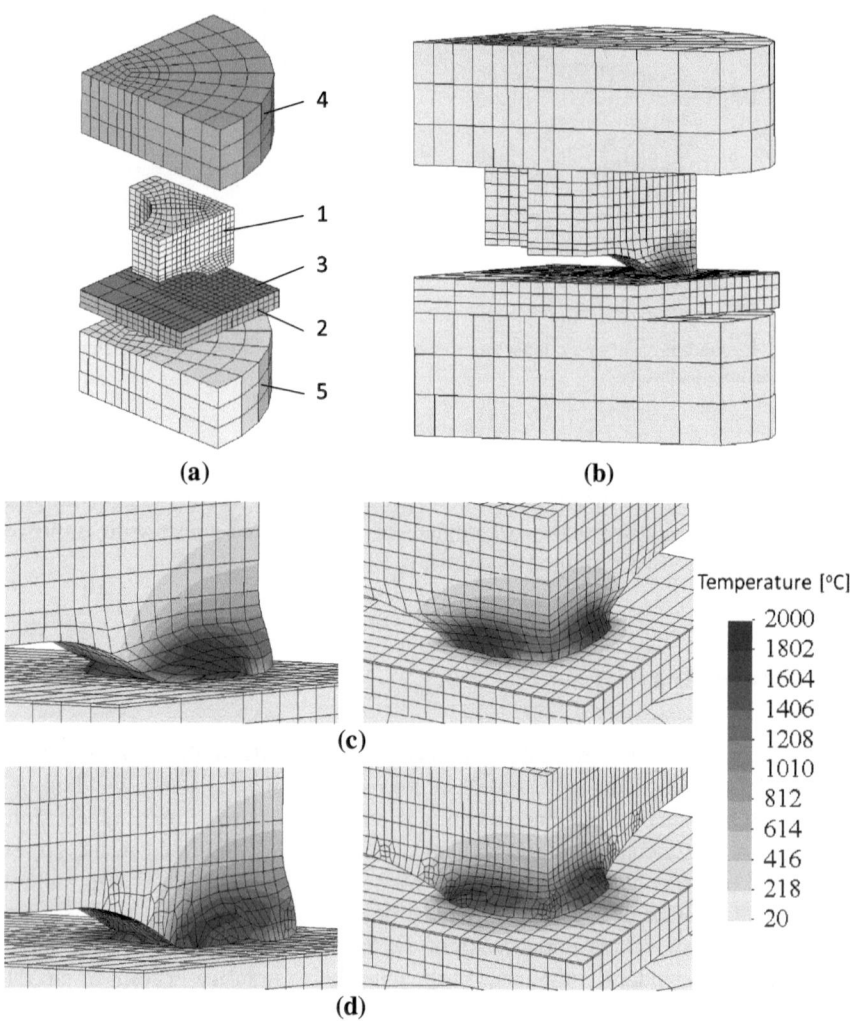

Fig. 5.7 Resistance welding of a square nut to a sheet. **a** Initial mesh of the multi-object finite element model. **b** Temperature distribution in original deformed mesh after 80 ms. **c** Temperature distribution in original deformed mesh after 100 ms. **d** Temperature distribution in the remeshed configuration after 100 ms, where remeshing took place after 80 ms

At the stage shown in Fig. 5.7b several elements in the bottom of the square nut (1) are flattened such that remeshing becomes necessary to carry on the simulation with high accuracy. However, the meshes in the remaining objects (2–5) are practically not distorted and, therefore, do not undergo remeshing. In other words, remeshing is only performed in the selected object (1). In case of remeshing, the surface mesh is extracted from the deformed geometry by splitting each of the surface quadrilaterals into two triangular elements. Hereafter follow the procedures outlined in Sect. 5.3 for the individual object.

5.4.1 Multi-Object Procedures and Tool Contact

The presence of multiple objects poses the necessity of paying special attention to ensure that penetration or gaps are avoided in regions where contact conditions prevail. The majority of nodes after remeshing will in general be located on the element faces of the previous distorted mesh but without coinciding with previous nodes. As a result, the surface of an object after remeshing will not be identical to the surface before remeshing, and therefore contact conditions are not guaranteed to be maintained unless the interfaces are planar. The solution is to reposition nodes of one object by orthogonal projection to an element face of another object if the orthogonal distance between them is less than a certain threshold tolerance. Additionally, all surface nodes of an object are tested for penetration into elements belonging to any other objects even if it exceeds the aforementioned tolerance.

A similar procedure is implemented for maintaining contact conditions between an object and a rigid tool with the constraint that only nodes of the object can be moved.

5.4.2 Transfer of History Dependent Variables

An additional step is necessary to complete the remeshing. The history dependent field variables, such as strain, damage, current density and temperature, need to be transferred from the old to the new mesh. This requires the evaluation of the nodal values of these quantities in the old mesh.

Averaging by weighted volumes of surrounding Gauss points is frequently applied, but it is possible to compute better nodal values extrapolated from Gauss points by applying a recovery technique based on least square fitting. The application of least square fitting requires the minimization of the following functional, I,

$$I = \int \sum_k \left[\sum_i N_i f_i - c_k \right]^2 dV_k \qquad (5.4.2.1)$$

where c_k is the known value of the time-integrated field variable at the centre Gauss point of element k, f_i is the nodal quantity to be determined and N_i is the conventional shape function of node i. Details of the procedure are comprehensively described by Martins et al. [20] and Fernandes and Martins [17].

Detailed views of the mesh and transfer of field variables in the example of the square nut to sheet welding case are provided in Fig. 5.8. A large number of elements have been applied to capture the details of the leg of the square nut (see also Fig. 5.7c, d). However, because the remaining part of the square nut has no or little deformation, the mesh density in this region is made lower in order to reduce the overall number of elements. The resulting mesh is shown in Fig. 5.8, where the entire square nut is shown in the lower figures and the details near a

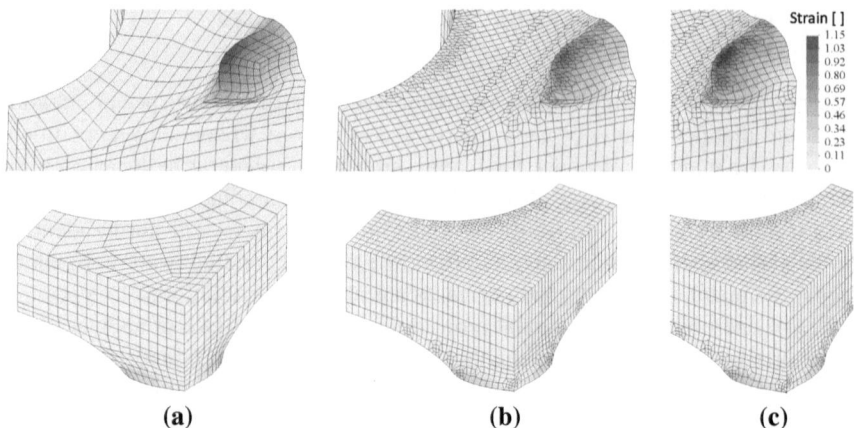

Fig. 5.8 Mesh and remesh details of the square nut after 80 ms. Transfer of field variables is performed for the effective strain. **a** Distribution of effective strain in the original mesh. **b** Distribution of effective strain after remeshing based on the averaged values of surrounding Gauss points weighted by element volumes. **c** Distribution of effective strain after remeshing based on least square fitting according to (5.4.2.1)

leg are shown in the upper figures. The overall number of elements is raised by a factor of 2.5 due to remeshing and resulting refinement in order to capture all the technological relevant local details.

The transfer of field variables (here exemplified by the effective strain) is performed from the original mesh in Fig. 5.8a to the new mesh in Fig. 5.8b, c. In Fig. 5.8b the transfer is accomplished by averaging of neighboring Gauss point values by weighted volumes, whereas the least square method according to (5.4.2.1) has been applied in Fig. 5.8c. As observed, the peak values (compare the dark color) are kept better when the transfer is performed by least square fitting than by volume weighted averaging.

References

1. Coupez T, Soyris N, Chenot JL (1991) 3-D finite element modelling of the forging process with automatic remeshing. J Mater Process Technol 27(1–3):119–133
2. Tekkaya AE, Martins PAF (2009) Accuracy, reliability and validity of finite element analysis in metal forming: a user's perspective. Eng Comput 26:1026–1055
3. Benzley SE, Perry E, Merkley K, Clark B (1995) A comparison of all hexagonal and all tetrahedral finite element meshes for elastic and elasto-plastic analysis. In: Proceedings of the 4th international meshing roundtable, pp 179–191
4. Kraft P (1999) Automatic remeshing with hexahedral elements: problems, solutions and applications. In: Proceedings of the 8th international meshing roundtable, South Lake Tahoe, pp 357–367
5. Schneiders R, Bünten R (1995) Automatic generation of hexahedral finite element meshes. Comput Aided Geom Des 12:693–707

6. Santos A, Makinouchi A (1995) Contact strategies to deal with different tool descriptions in static explicit FEM for 3-D sheet metal forming simulations. J Mater Process Technol 50:277–291
7. Boussetta R, Coupez T, Fourment L (2006) Adaptive remeshing based on a posteriori error estimation for forging simulation. Comput Methods Appl Mech Eng 195:6626–6645
8. Behrens BA, Kerkeling J (2011) Pierced forgings: tool development for a combined single step process. Prod Eng—Res Dev 5:201–207
9. Fernandes JLM, Alves LM, Martins PAF (2012) Forming tubular hexahedral screws—process development by means of a combined finite element-boundary element approach. Eng Anal Boundary Elem 36:1082–1091
10. Rodrigues JMC, Martins PAF (1998) Coupled thermo-mechanical analysis of metal-forming processes through a combined finite element-boundary element approach. Int J Numer Meth Eng 42:631–645
11. Zienkiewicz OC, Phillips DV (1971) An automatic mesh generation scheme for plane and curved surfaces by isoparametric coordinates. Int J Numer Meth Eng 3:519–528
12. Martins PAF, Barata Marques MJM (1992) Model3—a three-dimensional mesh generator. Comput Struct 42(4):511–529
13. Schneiders R, Oberschelp W, Kopp R, Becker M (1992) New and effective remeshing scheme for the simulation of metal forming processes. Eng Comput 8:163–176
14. Zhu J, Gotoh M (1999) An automated process for 3D hexahedral mesh regeneration in metal forming. Comput Mech 2:373–385
15. Karadogan C, Tekkaya AE (2001) Geometry defeaturing and surface relaxation algorithms for all-hexahedral remeshing. Proceedings of the Seventh NUMIFORM Conference, Toyohashi, pp 161–166
16. Kwak DY, Im YT (2002) Remeshing for metal forming simulations—part II: three-dimensional hexahedral mesh generation. Int J Numer Meth Eng 53:2501–2528
17. Fernandes JLM, Martins PAF (2007) All-hexahedral remeshing for the finite element analysis of metal forming processes. Finite Elem Anal Des 43:666–679
18. Nielsen CV, Fernandes JLM, Martins PAF (2013) All-hexahedral meshing and remeshing for multi-object manufacturing applications. Submitted for publication in international journal
19. O'Rourke J (1998) Computational Geometry in C. Cambridge University Press, Cambridge
20. Martins PAF, Marmelo JCP, Rodrigues JMC, Barata Marques MJM (1994) Plarmsh3—a three dimensional program for remeshing in metal forming. Comput Struct 53:1153–1166

Chapter 6
Parallelization of Equation Solvers

When solving large finite element problems, solution time becomes a factor which cannot be ignored. It is among the concerns when considering modeling in three dimensions instead of two dimensions. Different approaches are available to reduce the computational cost. Decomposition of a finite element domain into subdomains allows naturally for parallel computation of the subdomains to save overall computation time; see e.g. El-Sayed and Hsiung [1]. Interface nodes between substructures couple the substructure solutions, and thus communication between the processors are needed. In order to keep the amount of interface nodes minimal, Farhat [2] and Al-Nasra and Nguyen [3] have proposed algorithms for optimal decompositions. Another way of saving computation time is to apply faster solution techniques to solve the system of equations. This can be done either by solving iteratively, sequentially or in parallel, or by parallelizing the equation solver, such that it remains a direct solver.

6.1 Strategies of Solution Techniques

In iterative solvers, the solution is found iteratively to satisfy the equation system to within a specified tolerance. This is faster than directly solving the equation system as long as the rate of convergence is fast enough. Lanczos [4] and Hestenes and Stiefel [5] have e.g. proposed conjugate gradient (CG) iterative solvers, which were later improved by preconditioning, see Meijerink and van der Vorst [6], where a matrix is multiplied to each side of the system to precondition the system and thereby improving the convergence behavior.

The drawback of the iterative solving is that accuracy is lost compared to direct solving. The accuracy depends on the threshold value used for accepting the solution, and a compromise between accuracy and computation speed is necessary. The small inaccuracies accumulate and may result in poor satisfaction of boundary conditions, and symmetries may not be exactly obeyed (for instance, a zero

C. V. Nielsen et al., *Modeling of Thermo-Electro-Mechanical Manufacturing Processes*, SpringerBriefs in Applied Sciences and Technology, DOI: 10.1007/978-1-4471-4643-8_6, © The Author(s) 2013

displacement associated with a symmetry condition may be computed as a very small non-zero displacement creating problems in the overall modeling accuracy). In problems involving contact it may, for larger threshold values, also disturb the contact algorithms, eventually leading to penetration. Iterative solvers have also been reported unstable when dealing with ill-conditioned equation systems, whereas direct solvers are more robust, cf. Farhat and Wilson [7]. Due to the highly ill-conditioned systems dealt with in the present finite element implementation (the irreducible flow formulation with penalty contact), direct solvers are preferred and therefore parallelization of the iterative solver will not be considered, although it should be considered for other finite element implementations.

Parallelizing direct solvers is another way of saving computation time. The parallelization itself is considered more tedious, but once it is done, the time savings are easily obtained, and the accuracy is maintained to precision comparable to the sequential direct solver. Applying a parallel direct solver also diminishes the need for decomposition, although they can go together. Diminishing of this necessity entails that the parallel solver can be directly applied to any problem. Parallelization can be applied for local memory processors as well as for shared memory processors, where the first typically is applied to a cluster of multiple computers, whereas the latter typically would be one computer with multiple threads.

6.2 Parallel Skyline Solver

This section deals with parallelization of a skyline solver and follows the algorithms and computer implementation that was originally developed by Nielsen and Martins [8]. The skyline format of the system matrices is chosen because of the large sparsity typical for finite element models. Alternative compressed sparse row storage formats would also be relevant. When finite element programs are transferred to the industry, they are increasingly often intended for execution on standard PC's, which nowadays are equipped with several cores and threads with shared memory. It is therefore an obvious request that the programs can utilize all the threads to reduce the computational time.

The skyline solver is parallelized by OpenMP instructions in a FORTRAN implementation with details and source code provided by Nielsen and Martins [8] and with source code reproduced in Appendix A of this book.

The parallel skyline solver is easily implemented into existing finite element codes as only the call to the skyline solver has to be replaced by a call to the presented solver. The requested inputs are the stiffness matrix in skyline storage format together with the corresponding pointers to the diagonal positions, the right hand side, the number of equations and the number of threads to be utilized. Before this solver, Farhat and Wilson [7] published a parallel skyline solver programmed in Force, and Synn and Fulton [9] have proposed procedures to predict the performance of parallel skyline solving.

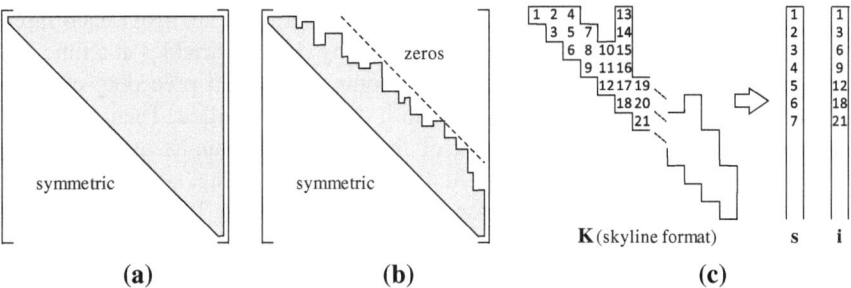

Fig. 6.1 System matrix in skyline storage format. **a** System matrix storage by utilizing symmetry only. **b** Skyline format by omitting zeros. The *dashed line* by the tallest skyline indicates the storage in banded format. **c** Format of skyline vector **s** and index vector **i** based on the original system matrix **K**. Numbers correspond to position in the skyline vector

A regular system of equations, like $\mathbf{Kv} = \mathbf{f}$, is considered, where $\mathbf{K}$ is a symmetric $n \times n$ matrix and $\mathbf{v}$ and $\mathbf{f}$ are $n \times 1$ vectors containing the unknowns and the right hand side, respectively. Due to symmetry of the system matrix, only half of the matrix needs to be built and stored (slightly more than half due to storage of all diagonal positions). Furthermore, since most finite element systems are sparse and by proper node numbering have many zeros far from the diagonal, a skyline format as depicted in Fig. 6.1 is adopted. Omitting all zeros above the skyline reduces the storage and later the solution time significantly. Zeros may still exist below the skyline as the skyline encloses all non-zero positions. In skyline format, the system matrix is typically stored in a one-dimensional vector **s** with an additional index vector **i** pointing to the diagonal positions. This is illustrated in Fig. 6.1c up to the seventh column. The size of the skyline vector is the number of positions under the skyline. The size of the index vector equals the number of rows or columns n. Then, it follows that the size of the skyline vector is i_n, since the last diagonal is the last position in the skyline vector.

The solution of the equation system is commonly performed by Gauss elimination with column reduction, which is divided into the following three steps:

- Factorization of system matrix and reduction of right hand side (this step is performed column by column, thereby being "with column reduction").
- Division of right hand side by system matrix diagonals.
- Backward substitution.

The factorization of the system matrix and reduction of the right hand side is parallelized, whereas the division of the right hand side by the system matrix diagonals as well as the backward substitution are left sequential because the time spent on these tasks are marginal compared to the factorization and reduction.

The parallelization is column based in the sense that each thread is assigned a column to process, and when finishing one column assigned the next unprocessed column. The columns, however, cannot be processed independently, implying that communication between the threads is necessary. This is accomplished through the shared memory by updating the relevant variables from each thread while making sure that only one thread is updating certain variables at a time. The complete processing of a column requires completion of all preceding columns, but partial processing can be initiated even if this is not fulfilled. Then, while performing the partial processing, more of the preceding columns may have been fully processed in other threads, and in that case the remaining, or yet another partition, of the column can be processed. This procedure may lead to waiting time in each thread while dependent variables are being processed in other threads, especially when the differences in the skyline heights are large, corresponding to increased unevenness of the skyline profile in Fig. 6.1b.

6.3 Comparison of Skyline Solver with Other Solvers

The parallel skyline solver is compared with a band solver and an iterative solver. The band solver is a direct solver as the skyline solver, but it works on a system matrix stored in band form shown in Fig. 6.1b by the dashed line and many zeros are therefore stored and processed compared to the skyline storage format decreasing the overall efficiency. The iterative solver included in the comparison is based on the conjugate gradient method with preconditioning; see more details provided by Fernandes and Martins [10]. All simulations are performed on a Dell Optiplex 980 desktop with an Intel(R) Core(TM) i7-860 processor with four cores and eight threads. It has 8 GB RAM, 8 MB cache and a clock frequency of 2.8 GHz. The system is 64-bit, but the program is running in 32-bit. The operating system is Windows 7. In order to keep the computer under the same global workload when testing the solution speed, all eight threads have been active during all simulations. When testing solution time using N threads, the remaining $8 - N$ threads have been running similar dummy simulations.

Figure 6.2a shows the test case used in the comparison of different solvers. The test is simple upsetting of a cube between two flat parallel platens. Two of the cube faces have prescribed symmetry, and contact between the cube and the tools is frictionless. The cube with dimensions $10 \times 10 \times 10$ mm^3 is compressed to half height through 100 simulation steps of $\Delta t = 0.05$ s with a velocity $v = 1$ mm/s. The cube with material described by the flow stress curve $\sigma = 180.65\varepsilon^{0.183}$ MPa is discretized by e^3 8-node isoparametric elements of equal initial size. This discretization implies $(e + 1)^3$ nodes and $3 (e + 1)^3$ degrees of freedom with three unknown velocity components per node.

Figure 6.2b shows the solution time as function of the number of degrees of freedom. The solution time is normalized by the solution time of the parallel skyline solver using eight threads. As expected, the band solver is much slower than the other solvers, and having the other solvers available, the band solver becomes outdated. Among the skyline solvers, the solution time is ideally halved when going from sequential (one thread) to two threads, from two to four threads, and from four to eight threads. The solution time is not completely halved since

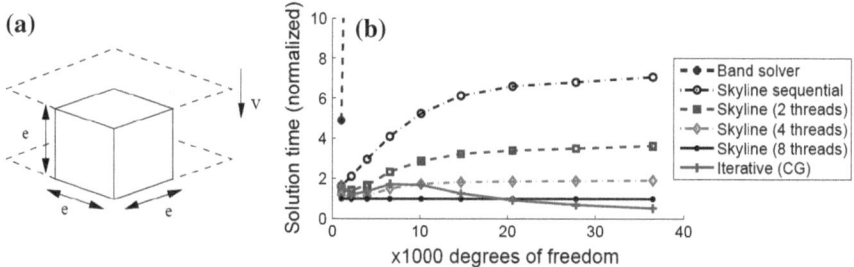

Fig. 6.2 Comparison of band solver, iterative solver and parallel skyline solver. **a** Simple upsetting test case. **b** Normalized solution time as function of degrees of freedom. The solution time is normalized by the solution time of the skyline solver using eight threads

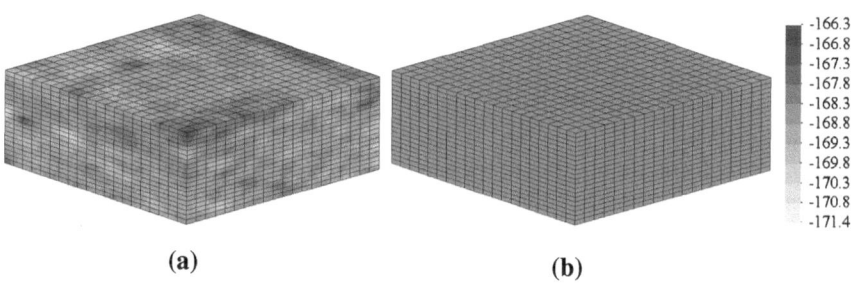

Fig. 6.3 Vertical component of the stress field in the cube compression example for 20 elements along each side. **a** Solution by iterative solver. **b** Solution by direct skyline solver (any number of threads). Common scale bar takes the minimum and maximum values according to (**a**)

the program is not 100 % parallel due to the waiting time described in the end of Sect. 6.2, other tasks than equation solving, and due to overhead. The iterative solver has solution time comparable with the skyline solver. Comparing to the parallel skyline solver using eight threads, the iterative solver is slower below approximately 20,000 degrees of freedom, and above it is faster. When using fewer threads in the skyline solver, this separation number of degrees of freedom is smaller. On the other hand, if more threads were available, the parallel skyline solver would be faster than the iterative solver at even larger number of degrees of freedom.

When the solution times of the iterative solver and the parallel skyline solver are in the same range, the iterative solver has the benefit that other threads are still available for other computations. However, the skyline solver has the benefit of being a direct solver implying better accuracy than the iterative. Figure 6.3 shows an example of the accuracy differences between the solvers. The vertical stress component is shown on the cube after compression to half height. The resulting stress distribution when applying the iterative solver varies as shown in Fig. 6.3a between -166.3 MPa and -171.4 MPa, whereas the distribution when applying the direct skyline solver is uniform with a value of -168.8 MPa. The deviations in the result from the iterative solver are $-1.48\,\%$ and $1.54\,\%$ relative to the result from the direct solver. Observations of the iterative solver showed that the solution did not converge in any of the 100 steps. Decreasing the tolerance for convergence would therefore not have an effect. Instead, the limit on the number of iterations was removed to ensure convergence (if possible). The average number of iterations was increased about 1.20 times, and the solution of the equation system converged in all steps. The above deviations reduced to $-0.178\,\%$ and $0.237\,\%$. However, due to the increased number of iterations, the solution time increased about 13 %. The longer solution time for the iterative solver makes it less attractive than it appears in Fig. 6.2b. For a more ill-conditioned system of equations (e.g. due to rigid zones or contact between deformable bodies), the increase of iterations and solution time would be even more.

The differences between the results in Fig. 6.3, even when the iterative solver converges, become crucial when analyzing more complex geometries including contact between deformable bodies. On top of accuracy problems, the iterative solver may become unstable when dealing with ill-conditioned equation systems; see Farhat and Wilson [7] and Fernandes and Martins [10]. Ill-conditioned equation systems are likely to appear when penalty methods are applied as in the present computer program.The skyline solver has therefore been adopted as the standard solver, and after the parallelization, the solution time is not further minimized by an iterative solver for the majority of system sizes dealt with.

6.4 Performance Evaluation of Parallel Skyline Solver

Speed-up is evaluated based on the compression of a cube to half height presented in Sect. 6.3. The speed-up (ratio of the solution time on one thread to the solution time on N threads—ideally equal to N) is shown in Fig. 6.4 as function of degrees of freedom. The finite element program is not entirely parallel, so the speed-up is less than the ideal. Part of the program is still sequential, since only the equation solver of the main system of equations has been parallelized, and in addition heading (physical communication to and between the threads) takes time. As the system size (degrees of freedom) increases, relatively more time is necessary to solve the equation system, which means that the fraction of the code running in parallel becomes relatively larger. This results in the larger speed-up seen in Fig. 6.4 at increasing number of degrees of freedom. It is also seen in the figure that the speed-up is largest for the smaller number of threads. This is a result of increased heading time and increased waiting time between threads when more threads are used, but it is also a result of a relatively smaller time fraction being parallel, simply because the amount of solution time with more threads is less compared to the overall time.

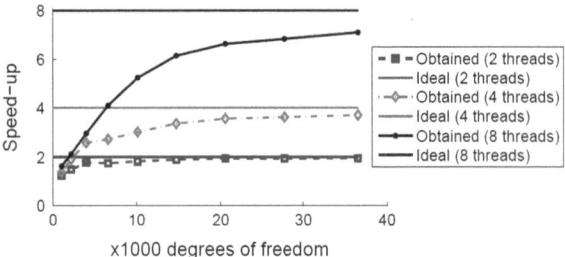

Fig. 6.4 Speed-up for 2, 4 and 8 threads as function of degrees of freedom

Amdahl's law (originating from Amdahl [11]) is estimating the speed-up $\tilde{\sigma}$ for a certain number of threads N based on the fraction P of the program being parallel according to

$$\tilde{\sigma} = \frac{1}{(1 - P) + \frac{P}{N}} \qquad (6.4.1)$$

where $1 - P$ is the sequential contribution and $\frac{P}{N}$ is the parallel contribution. Rearranging allows the estimation of the parallel fraction from the actual speed-up.

For the larger system sizes (here 36,501 degrees of freedom), the parallel fraction is estimated to approximately 97.5 % for all number of threads. Note that this is of the entire program, not only the skyline solver, which means that the CPU time taken in the rest of the program is negligible. Insertion into (6.4.1) with the number of threads going to infinity shows that this parallel fraction corresponds to a maximum achievable speed-up of 40. Amdahl's law can also estimate the speed-up for a standard PC with 16 threads to 11.6. Dreaming further to reach e.g. 32 and 64 threads in standard PC's, the estimated speed-ups are 18.0 and 24.9, respectively.

6.4.1 Evaluation by a Resistance Welding Case

The parallel skyline solver is also tested for an industrial case by evaluating the solver in simulation of resistance welding with different number of threads. The welding case is shown in Fig. 6.5a and consists of two AISI 1008 steel alloy sheets of 1 mm thickness that are spot welded between two copper alloy electrodes with tip diameter ∅6 mm.

The electrode center axes are placed in a distance 13 mm to three of the sheet edges, but only 4 mm from the fourth edge. Total simulated process time is 340 ms. The electrode force is raised linearly to 3 kN within 20 ms and kept constant hereafter. AC current is applied after 40 ms, lasting 200 ms at a level of 8 kA RMS with a conduction angle of 80 %. After the current is turned off, the electrode forces are kept for additionally 100 ms while the weld nugget solidifies.

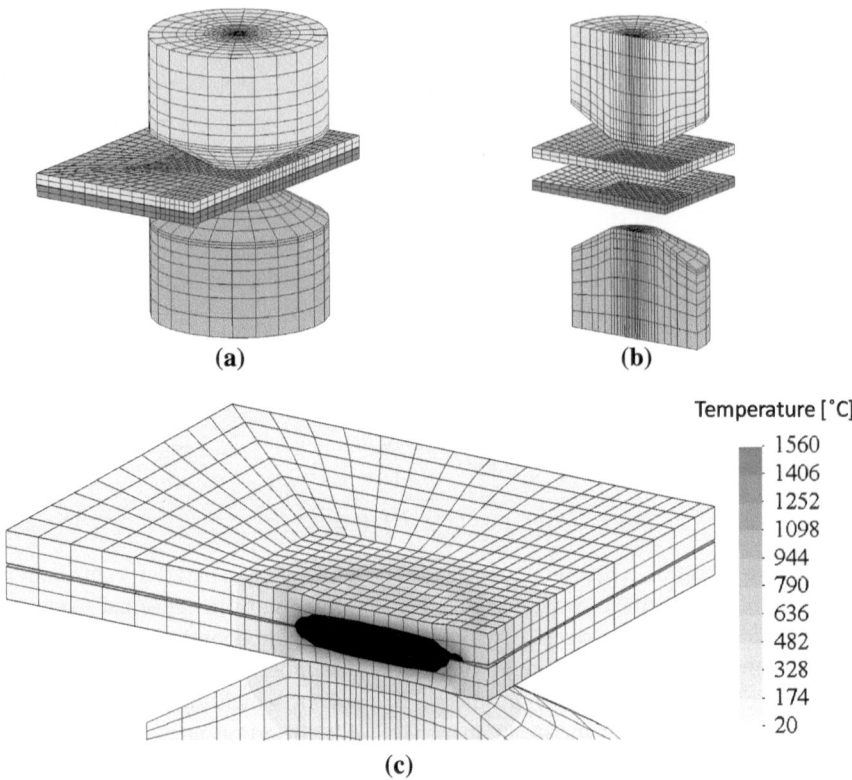

Temperature [°C]

- 1560
- 1406
- 1252
- 1098
- 944
- 790
- 636
- 482
- 328
- 174
- 20

(a)   (b)

(c)

Fig. 6.5 Resistance spot welding test case. **a** Arrangement of electrodes and sheets for testing spot welding near an edge. **b** Applied mesh using symmetry. Total number of nodes is 7,666. **c** Temperature field (shown without upper electrode) in the end of the weld time

The mesh shown in Fig. 6.5b consists of 7,666 nodes giving rise to 22,998 degrees of freedom in the mechanical model and 7,666 degrees of freedom in the electrical and thermal models. Figure 6.5c shows the resulting temperature field after the applied welding time. The spot seems almost axisymmetric showing that the chosen distance to the edge may not be a problem. However, this is without analysis of splash, which may be determining. Due to less material on the edge side, the temperature decreases slower near the edge and this asymmetric cooling may result in a microstructure and residual stress distribution that the welding engineer has to be aware of.

The solution times in the welding case are shown in Fig. 6.6 for two approaches to node numbering optimization. The approach in Fig. 6.6a is optimization of node numbering without consideration of contact between the objects shown individually in Fig. 6.5b. The approach in Fig. 6.6b is optimization of node numbering including information of initial contact, i.e. the contact arising when the objects are just brought vertically together. Solution times are shown as function

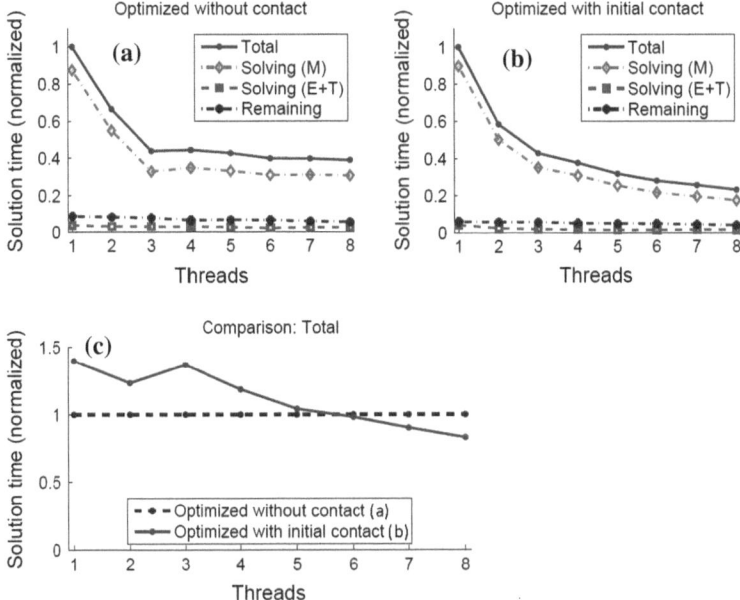

Fig. 6.6 Solution time of the entire solution, the pure equation solving in the mechanical model (M), and the combined pure equation solving in the electrical and thermal models (E + T). The solution time of the "remaining" is also included, which is the total subtracted the pure equation solving. **a** Normalized solution time as function of applied threads with node numbering optimization independent of contact. **b** Normalized solution time as function of applied threads with initial contact included in the node numbering optimization. **c** Comparison of total solution times in **a** and **b**

of applied threads, where, as in the above analyses, when applying N threads, the remaining $8 - N$ threads have been applied to a similar dummy simulation.

The solution times are shown for the total running time of the entire simulation as well as for the pure solution time of the equation system in the mechanical model and in the electrical and thermal models. These pure solution times are accumulated over the entire simulation. It is clear from the figures that the equation solving in the mechanical model is the main contributor to the total solution time. The combined solution time in the electrical and thermal models is much less, partly due to fewer iterations and in particular due to the smaller system size (7,666 degrees of freedom compared to 22,998 degrees of freedom in the mechanical model). The figures also include the remaining time spent in the simulation, i.e. the total time subtracted the pure solution time in the main equation systems. Thus, the remaining solution time is a sum of setting up the equation systems, searching for and evaluating contact, updating variables before time stepping, etc.The overall solution time decreases with increasing number of applied threads. This is mainly accommodated by the shorter time spent in the pure solution of the mechanical equation system. The time spent in pure solution of the electrical and thermal equation systems decreases only little, and the time spent on remaining tasks should be unchanged, since it is not parallelized.

An interesting difference is observed between the two approaches to the node numbering optimization. When the optimization is performed without information of the contact, the solution time does not decrease noticeable when applying more than three threads (Fig. 6.6a). On the other hand, the solution time decreases remarkably over the whole range of applied threads when node numbering optimization includes information on initial contact (Fig. 6.6b). The reason for this difference is explained by Nielsen and Martins [8] to stem from peaks in the skyline heights due to contact. When contact is not included in the node numbering optimization, the skyline profile becomes relatively even with low skyline heights, but when contact is detected, the skyline has to be expanded as discussed in relation to Fig. 4.4 resulting in peaks of the skyline profile. The differences in skyline heights reduce the speed-up above a certain number of threads. On the other hand, when initial contact is included in the optimization of node numbering, a more even skyline profile is achieved after contact between the objects, but on the expense of an overall increase of skyline heights.

When it comes to speed-up, the above comparison shows that the node numbering optimization taking initial contact into account is clearly better than the optimization without contact information. Figure 6.6c compares the actual solution time of the two approaches for different numbers of applied threads. The solution times are normalized by the solution time of the optimization without contact information. The figure shows that the approach including initial contact (which has the better speed-up) is slower by a factor of 1.4 when using one thread. However, due to the better speed-up, it becomes faster when applying six or more threads. The reason for the slower solution when using few threads is that the initial contact is much more than the contact after separation of the sheets outside the weld zone, and therefore the optimized skyline according to the initial contact is not optimal throughout the entire solution.

Figure 6.6b, as well as the figures related to the cube compression example, proves that parallelization of skyline solvers is computationally efficient. Hereafter, it is up to a correct approach for the node numbering optimization to get the best use out of it. In the specific welding case, an improved strategy would be to start out with an optimized node numbering based on the initial contact, and then reoptimize the node numbering when the sheets have separated outside the weld zone.

References

1. El-Sayed MEM, Hsiung C-K (1990) Parallel finite element computation with separate substructures. Comput Struct 36(2):261–265
2. Farhat C (1988) A simple and efficient automatic fem domain decomposer. Comput Struct 28(5):579–602
3. Al-Nasra M, Nguyen DT (1992) An algorithm for domain decomposition in finite element analysis. Comput Struct 39(3/4):277–289

4. Lanczos C (1952) Solution of systems of linear equations by minimized iterations. J Res Nat Bur Stand 49(1):33–53
5. Hestenes MR, Stiefel E (1952) Methods of conjugate gradients for solving linear systems. J Res Nat Bur Stand 49(6):409–436
6. Meijerink JA, van der Vorst HA (1977) An iterative solution method for linear systems of which the coefficient matrix is a symmetric m-matrix. Math Comput 31(137):148–162
7. Farhat C, Wilson E (1987) A parallel active column equation solver. Comput Struct 28(2):289–304
8. Nielsen CV, Martins PAF (2013) Parallel skyline solver: implementation for shared memory on standard personal computers. Submitted for publication in International Journal
9. Synn SY, Fulton RE (1995) The performance prediction of a parallel skyline solver and its implementation for large scale structure analysis. Comput Syst Eng 6(3):275–284
10. Fernandes JLM, Martins PAF (2009) Robust and effective numerical strategies for the simulation of metal forming processes. J Manuf Technol Res 1(1/2):21–36
11. Amdahl GM (1967) Validity of the single processor approach to achieving large scale computing capabilities. In: AFIPS Joint Computer Conferences, pp 483–485

Chapter 7
Material, Friction and Contact Characterization

Awareness and understanding of the basic procedures to determine the flow stress, the frictional response and the electric and thermal contact resistances under different conditions of strain-rate and temperature are fundamental for improving the quality of data to be inserted in finite element computer programs. Because accuracy and reliability of numerical simulations are critically dependent on input data, the following sections will provide a brief overview of the most widespread experimental techniques that are utilized for material, friction and contact characterization.

7.1 Mechanical Properties at Room Temperature

From a metal forming point of view, the most important data for modeling material behavior is the flow curve because it characterizes strain-hardening and determines the force and work requirements of a process as well as the relative material flow. In case of cold forming, the flow curve should be available to strain levels above "1" for bulk metal forming, and up to "1" for sheet metal forming processes.

The compression test performed on solid cylinder specimens is one of the most widespread mechanical testing methods for determining the flow curve in the field of metal forming. The capability of evaluating material response to much larger strains than in tensile tests, due to the absence of necking, in conjunction with the aptitude to better emulate the operative conditions of real forming processes, such as forging, rolling and extrusion, which are carried out under high compressive loads, are seen as the main reasons for its extensive utilization.

The compression test is performed by axially pressing a solid cylinder specimen between two flat polished, well lubricated, parallel platens and the flow curve is determined by combining the experimental values of force and displacement. A variant of the compression test is utilizing Rastegaev specimens, see Lange [1] and illustration in Fig. 7.1a, to reduce friction towards the platens by having a reservoir for the lubricant. This reduces barreling effectively, but leads to errors in measuring the height of the specimens due to bending of the surrounding walls and

C. V. Nielsen et al., *Modeling of Thermo-Electro-Mechanical Manufacturing Processes*, SpringerBriefs in Applied Sciences and Technology, DOI: 10.1007/978-1-4471-4643-8_7, © The Author(s) 2013

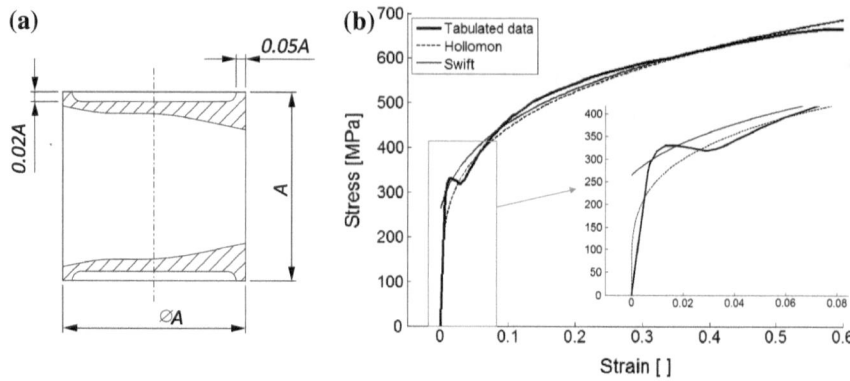

Fig. 7.1 Material testing by upsetting of Rastegaev specimens. **a** Geometry of the Rastegaev's compression test specimen. **b** Experimental stress–strain curve for a structural steel S235JR + AR and approximations by Hollomon and Swift curves. The size of the test specimens is defined by $A = 20$ mm

end faces not remaining plane. Figure 7.1b shows an example of a flow stress curve by tabulated data giving a best fit of the measured data, in this case best fit of six repetitions. Molykote DX paste was utilized as lubricant in the specific example.

The flow curve is in many cases approximated by fitting curves for easy description of the material, e.g., in finite element programs. Two typical approximations are shown in Fig. 7.1b by the Hollomon (7.1.1) and Swift (7.1.2) equations,

$$\sigma = C\varepsilon^n = 777\varepsilon^{0.243}[MPa] \qquad (7.1.1)$$

$$\sigma = C(B + \varepsilon)^n = 775(0.012 + \varepsilon)^{0.243}[MPa] \qquad (7.1.2)$$

where C is the flow stress at strain $\varepsilon = 1$ or $B + \varepsilon = 1$, n is the strain-hardening exponent and B corresponds to a pre-straining. Both fitted curves are representing the overall behavior, but details like the yield point phenomenon existing in low-carbon steels, see detail in Fig. 7.1b, cannot be captured by such approximations. Due to the additional parameter in terms of the pre-strain, the Swift equation provides a better approximation, but only taking the details near the yielding point in an average sense. To overcome this problem, computer programs not only include more sophisticated flow stress models (e.g. Johnson–Cook and Preston–Tonks–Wallace, among others) as they have the option of including tabulated data, such that the actual material response can be modeled.

The solid cylinder specimens utilized in the compression test are limited within the aspect ratio range $1 \leq h_0/d_0 \leq 3$ of the height h_0 to the diameter d_0, Gunasekera et al. [2] and Czichos et al. [3], though practically not exceeding $h_0/d_0 = 1.5$. The upper limit on the aspect ratio prevents failure by buckling or bending while the lower limit is commonly justified by the increased sensitivity to friction along the contact interface with compression platens (Alves et al. [4]), by technical difficulties to operate extensometers directly on the specimens, House [5], or not having enough displacement at all compared to the uncertainty of the

 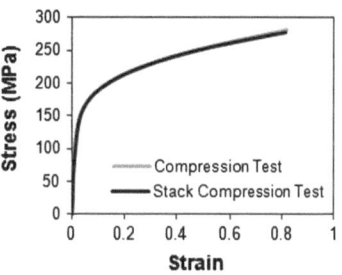

Fig. 7.2   Conventional and stack compression tests of Aluminum AA2011-O

measurement. This inhibits the utilization of the compression test for constructing the flow curve of materials available in form of sheets and plates.

As discussed by Alves et al. [4], the stack compression test proposed by Pawelski [6] is the best alternative experimental procedure for evaluating the flow curve of raw materials supplied in form of sheets and plates. The test makes use of circular discs that are cut out of the blanks and stacked to form a cylindrical specimen with an aspect ratio in the range of solid cylinders employed in the conventional compression test (Fig. 7.2).

As shown in Fig. 7.2 the stack compression test can be utilized for the construction of flow curves, although the procedure is not standardized. The resulting flow stress is nearly identical to that obtained by means of conventional compression tests.

However, it is worth noting that compression (as well as tensile) tests are performed under proportional loading while metalworking processes often involve non-proportional or cyclic loading. During non-proportional loading, the strain path influences the flow stress behavior as discussed by Huml and Lindegren [7] for cyclic loading and shown by Tekkaya and Martins [8] in finite element modeling of fullering with intermediate 90° turning of the specimen in-between two blows. The simulation was able to model the load-displacement response accurately in the first blow, but not as accurate in the second blow due to induced anisotropy. This is important when analyzing multi-stage processes with different loading paths in each stage because uniaxial material testing (under proportional loading) can be insufficient for accurate modeling of such cases.

Flow curves for a large number of materials can be found in Doege et al. [9].

7.2 Friction Characterization

Part of the characterization of frictional behavior is the recognition of levels of normal pressure and corresponding selection of friction model. Amonton–Coulomb's law,

$$\tau_f = \mu p \qquad\qquad (7.2.1)$$

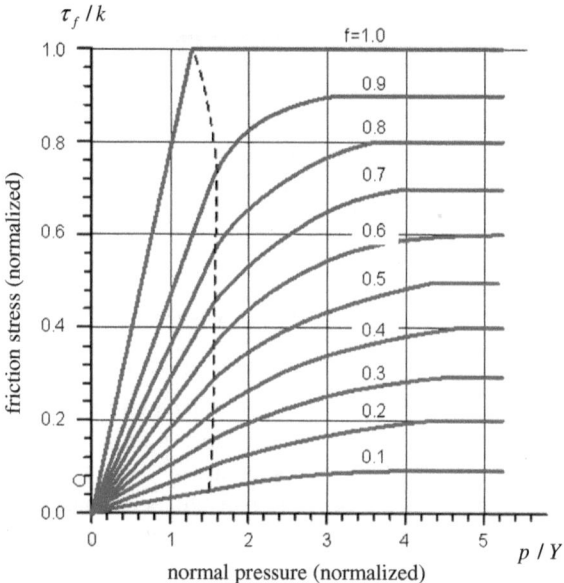

is prone to overestimate the friction in metal forming because of the high normal
pressures typically involved. On the other hand, the constant friction law,

$$\tau_f = mk \tag{7.2.2}$$

may also overestimate the friction in regions of low normal pressure because it
does not take into account the actual stress state. Wanheim and Bay [10] have proposed a general friction model resembling the two laws at low and high normal
pressures and providing a smooth transition in-between, see Fig. 7.3. The model,

$$\tau_f = f\alpha k \tag{7.2.3}$$

is based on slipline analysis calculating the ratio between real and apparent area of
contact α between a rough workpiece surface and a smooth tool surface assuming
the friction stress in the real area of contact τ_r to be constant and a fraction f of the
material shear flow stress k,

$$\tau_r = fk \tag{7.2.4}$$

where $0 \leq f \leq 1$. The curves are determined by discrete points but later put on
formula [11]. It should be pointed out that although the model in principle solves
the problem of describing friction in the entire interval from low to high normal
pressures, it does not account for bulk plastic deformation of the subsurface when
calculating the real contact area. This simplification implies underestimation of the
contact area and thus also friction.

As regards determination of friction data, μ, m or f, one of the well-known
standard tests is the ring compression test. If calibration curves are not available, they may be constructed by finite element simulation as shown in Fig. 7.4.

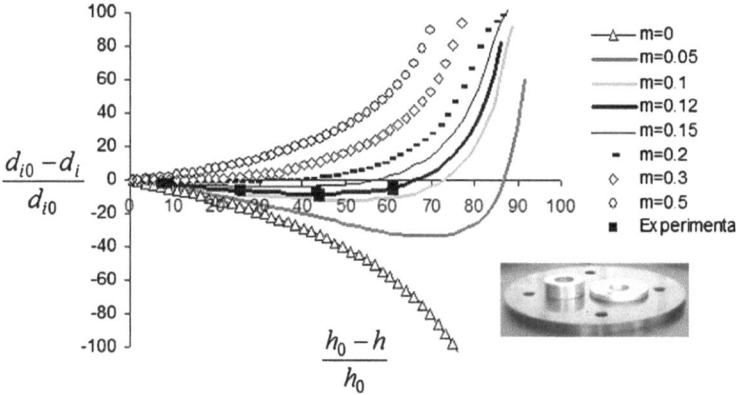

Fig. 7.4 Friction factor calibration curves obtained by finite element simulations under assumption of constant friction law and flow curve obtained from Aluminum AA1100-O. Experiments correspond to testing with lubricant Castrol Iloform PNW 124 mineral oil

The ring test, however, only supplies friction data for the given contact pressure and surface expansion valid for this test. It should furthermore be emphasized that the interface temperature during testing should be correctly emulated since viscosity of many metal forming lubricants is very sensitive to temperature.

Since modeling and quantification of friction by means of simple models such as Amonton–Coulomb's law and the law of constant friction is questionable, friction coefficients or factors are in many cases tuned by the users during the numerical simulation in order to provide good estimates of the forming loads and of the deformed shape of the workpiece.

7.3 Mechanical Properties at Elevated Temperatures

At elevated temperatures, e.g., in warm and hot forming processes or resistance welding, the flow curve is not only a function of the strain but also of the strain rate and temperature.

Figure 7.5 presents a set of flow curves obtained experimentally by upsetting Ø8 × 10 mm specimens between to flat parallel anvils at different temperatures and deformation rates. In this testing procedure, performed on Gleeble 1500 equipment, the temperatures in the specimens are controlled by sending high current pulses through the specimens to increase temperature. The temperature on the specimen surface is measured by a mounted thermocouple. The compression is performed in three intervals; an acceleration interval, a compression interval and an overtravel interval in order to obtain a strain rate during the compression interval as constant as possible. The specimen end faces are flat in contrast to the Rastegaev specimens in order to ensure proper contact to the anvils for the resistance heating. The friction is lowered by inserted graphite foils to minimize barreling. Additional

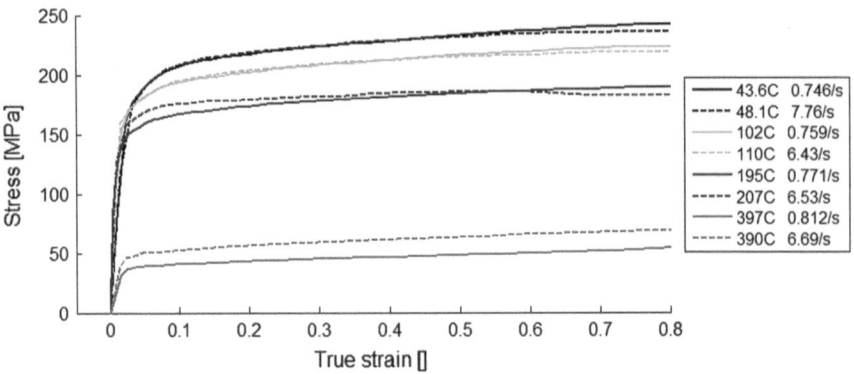

Fig. 7.5 Experimentally obtained flow curves for Aluminum AA6060-T6 at different temperatures and strain rates

corrections in the establishment of the flow curves are due to machine compliance and thermal expansion of the test specimens. This follows earlier work by Song et al. [12].

The material response represented by Fig. 7.5 is representative for many metals in terms of the lowered strength with increasing temperature. However, other responses can be identified by the testing procedure as e.g., blue brittleness in some steels, where the strength increases from room temperature to a level, say 400 °C, after which the strength decreases. The effect of strain rate is furthermore available from Fig. 7.5, showing little or no influence at lower temperatures, while at higher temperatures, the material has higher strength with increasing strain rate. The range of strain rates in the example is limited and sparse. Flow curves for a large number of materials at different temperatures and strain rates can be found in Doege et al. [9].

The need to perform material characterization for higher strain and strain rates than those currently attained requires the utilization of torsion testing machines, drop hammers, Hopkinson bar apparatus and inverse analysis. Viscous effects, such as viscoplastic behavior, are usually handled by a simplified approach of specifying the flow curves as a function of the equivalent plastic strain rate. However, it is important to notice that the associated constitutive equations are time independent.

7.4 Electrical Contact Properties

The electrical contact resistance across an interface between metals is difficult to predict and can vary significantly between batches or even from one weld to another. When two metal surfaces are brought in contact, only a fraction of the apparent area is in real contact. The load bearing area is formed on the surface

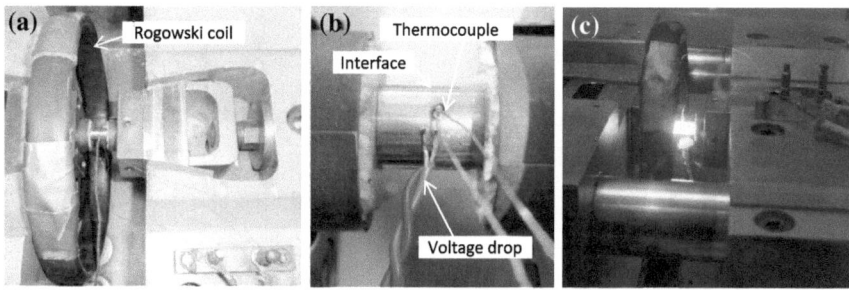

Fig. 7.6 Measurement of electrical contact resistance. **a** Test setup in Gleeble with a Rogowski coil to measure the applied current pulses. The test specimens are placed between two anvils applying a certain compression. **b** Close-up of the test specimens with mounted thermocouple for temperature measurement near the contact interface and mounted wires for measuring the voltage drop across the interface. **c** Example of testing at high temperature

asperities and increased with normal pressure. The current is therefore constricted with resulting increase of contact resistance. Additionally, surface films, coatings, oxides and contaminants will influence the contact resistance. Besides the dependency of the contact pressure, the contact resistance is also temperature dependent. The asperities become softer with increasing temperature and the resulting increased real contact area lowers the contact resistance. Increased contact pressure and/or temperature can cause break down and squeeze out of surface films, coatings, oxides and contaminants resulting in lowered contact resistance.

Figure 7.6 shows a test setup employed for characterization of the electrical contact resistance. Two cylindrical specimens are placed between the anvils in Gleeble equipment for the characterization of the interface between the two cylinders. The temperature is controlled by a thermocouple mounted close to the interface and high current pulses as described in Sect. 7.3. The contact pressure is controlled by movement of the anvils. A Rogowski coil is introduced as shown in Fig. 7.6a to measure the current of the applied pulses for heating the specimens, and the corresponding voltage drop over the interface is measured by mounted wires shown in Fig. 7.6b. Based on corresponding values of current and voltage drop, the resistance is given from Ohm's law. This follows earlier work by Song et al. [12].

The data pairs of current and voltage are selected at the time instants where the current peaks. This is to avoid the influence of induced electromotive force (emf) in the voltage measurement, which would otherwise lead to errors in the calculated resistance. The electromotive force is proportional to the first derivative of the current, and therefore supposed to vanish when the current peaks. It is furthermore proportional to the spanned area of the wires measuring the voltage drop, and the twisting of the wires seen in Fig. 7.6b is in order to minimize the spanned area.

Electrical bulk resistivity can be measured in a similar way by using only one specimen and typically increasing the length of the measured voltage drop. Obtained bulk resistivities are used to improve the calculation of contact resistance by subtraction of the resistance of the bulk material between the wires for

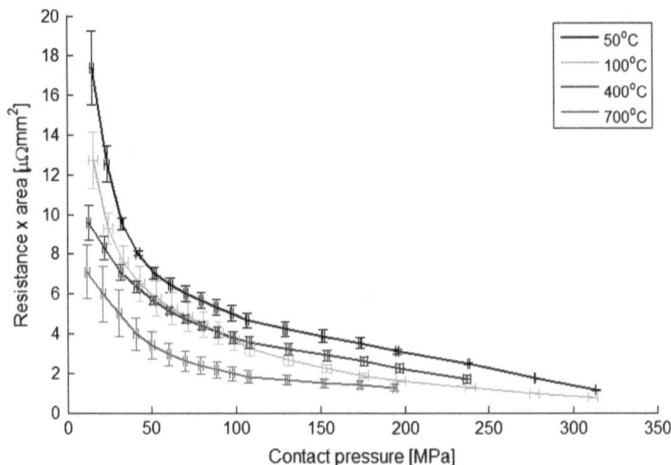

Fig. 7.7 Experimentally obtained contact resistance times contact area as function of contact pressure at different temperature levels for stainless steel AISI 316L

measurement of the voltage drop. Further corrections are due to the changed cross-sectional area and distance between the wires for voltage drop measurement stemming from the compression and thermal expansion.

An example of obtained contact resistance between two specimens of stainless steel AISI 316L with end faces prepared by turning is presented in Fig. 7.7 as function of contact pressure at different temperatures. The figure shows the typical behavior of decreasing contact resistance with increasing contact pressure and temperature. The contact resistance is represented by the product of the contact resistance and the contact area in order to present the data independent of the contact area, which changes during testing. The relation to the finite element modeling is obtained through

$$\rho_c l_c = R_c A_c \tag{7.4.1}$$

where $R_c A_c$ (the product of contact resistance and contact area) is directly the presented curves and l_c is the thickness of the contact layer of elements introduced in the simulations as interface layers. Once the thickness of the layer has been decided, the contact resistivity ρ_c is available for input to the simulation. The following model [13] is applied for the electrical contact resistivity between two materials in the numerical simulations,

$$\rho_c = \frac{3\sigma_{soft}}{\sigma_n}\left(\frac{\rho_1 + \rho_2}{2} + \rho_{contaminants}\right) \tag{7.4.2}$$

where σ_{soft} is the flow stress of the softer material in contact, σ_n is the contact normal pressure, ρ_1 and ρ_2 are the electrical bulk resistivities of the two contacting materials and $\rho_{contaminants}$ is the resistivity stemming from surface contaminants such as oxides, surface films and dirt. The model scales the contact resistivity through the fraction in front of the parenthesis according to the ratio of real contact area to the apparent contact area based on the theory by Bowden and Tabor [14]. The parenthesis consists of the average bulk resistivity plus the term $\rho_{contaminants}$ stemming from the actual surface condition. This term is used to scale the model according to the experimental curves (7.4.1) and Fig. 7.7.

References

1. Lange K (1984) Umformtechnik I. Handbuch für Industrie und Wissenschaft, 2nd edn. Springer-Verlag, Berlin (In German)
2. Gunasekera J, Chitty E, Kiridena V (1989) Analytical and physical modelling of the buckling behavior of high aspect ratio billets. Ann CIRP 38:249–252
3. Czichos H, Saito T, Smith L (2006) Handbook of materials measurements methods. Springer, Berlin
4. Alves LM, Nielsen CV, Martins PAF (2011) Revisiting the fundamentals and capabilities of the stack compression test. Exp Mech 51:1565–1572
5. House JW (2000) Testing machines and strain sensors in mechanical testing and evaluation. ASM International, Materials Park, pp 225–227
6. Pawelski O (1967) Über das stauchen von holzylindern und seine eignung zur bestimmung der formänderungsfestigkeit dünner bleche. Arch Eisenhüttenwes 38:437–442 (In German)
7. Huml P, Lindegren M (1992) Properties of cold-formed metal products. Ann CIRP 41(1):267–270
8. Tekkaya AE, Martins PAF (2009) Accuracy, reliability and validity of finite element analysis in metal forming: a user's perspective. Eng Comput 26(8):1026–1055
9. Doege E, Meyer-Nolkemper H, Saeed I (1986) Fliesskurvenatlas metallischer Werkstoffe. Hanser Verlag, München Wien, ISBN 3-446-14427-7 (In German)
10. Wanheim T, Bay N (1978) A model for friction in metal forming processes. Ann CIRP 27(1):189–194
11. Bay N (1987) Friction stress and normal stress in bulk metal–forming processes. J Mech Work Technol 14:203–223
12. Song Q, Zhang W, Bay N (2005) An experimental study determines the electrical contact resistance in resistance welding. Weld J 84(5):73s–76s
13. Zhang W (2003) Design and implementation of software for resistance welding process simulations. Trans J Mater Manuf 112(5):556–564
14. Bowden FP, Tabor D (1950) The fabrication and lubrication of solids. Oxford University Press, Oxford

Chapter 8
Applications

Accuracy, reliability and validity of the coupled finite element flow formulation are evaluated by performing numerical simulations of industrial manufacturing processes. Emphasis is put on joining technologies by tube forming and resistance welding due to its importance for assembling individual components together in complete and useful end products and also due to the fact that selected applications deal with state-of-the-art engineering concepts that are not commonly available in the open literature. Several of the presented examples are industrial cases.

8.1 Mechanical Joining of Tubes

Conventional tube branching technology provides joining of tubes by means of tee connections and is widely utilized in plumbing, air conditioning, refrigeration, process piping and lightweight structures, among other applications. There are several different joining methods and each has its own advantages and disadvantages (Fig. 8.1).

The most well-known types are based on commercially available tee fittings, saddle adapters and weld-o-lets for standard geometries and materials, such as carbon steel, stainless steel, copper and polyethylene, among other thermoplastics (Fig. 8.1a–c). A standard tee fitting (Fig. 8.1a) has three welds; two in the main tube and one in the branch tube. Saddle adapters or weld-o-lets (Fig. 8.1b, c) also need to be brazed or welded to the main tube over a pre-cut hole and the attachment to the branch tube is made through a weld or a threaded connection.

Because joining tubes by means of standard tee fittings or commercially available saddle adapters is not an appropriate technology for obtaining connections with non-standard geometries, there are alternative tube branching methods that take advantage of the user's ability to fabricate own connections (Fig. 8.1d–f). The nozzle-weld is the most commonly fabricated tee connection but other solutions based, for instance, on spin-forming are also frequently employed.

C. V. Nielsen et al., *Modeling of Thermo-Electro-Mechanical Manufacturing Processes*, SpringerBriefs in Applied Sciences and Technology, DOI: 10.1007/978-1-4471-4643-8_8, © The Author(s) 2013

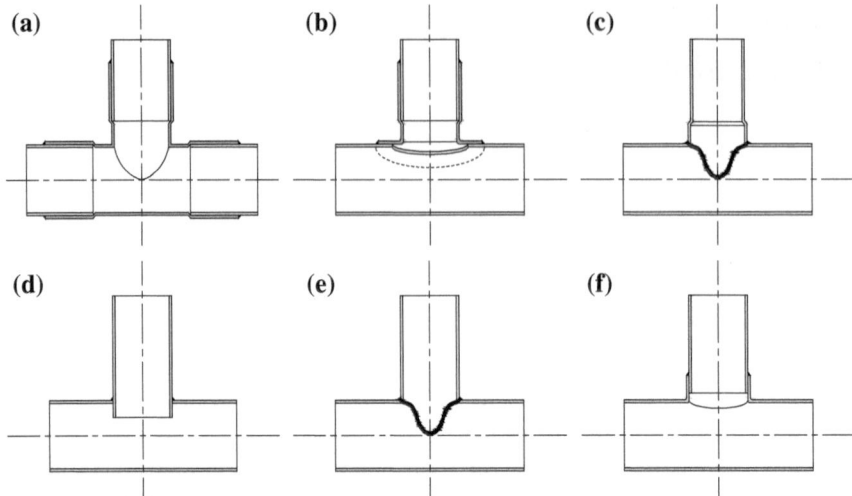

Fig. 8.1 Conventional tube branching methods by means of **a** tee fittings, **b** saddle adapters, **c** weld-o-lets, **d** nozzle-welds, **e** nozzle-welds (by shaping a contoured end in the branch tube) and **f** spin-forming

Nozzle-weld connections (Fig. 8.1d, e) require cutting a hole in the main tube, shaping a contoured end in the branch tube to match the diameter of the main tube and welding along the contour. In some cases the procedure is simplified by letting the branch tube to be inserted into the hole and just welding along the contour resulting from the intersection between the tubes.

Spin-forming (Fig. 8.1f) also requires cutting a hole in the main tube. The difference is that material around that hole is subsequently shaped into a tee fitting where the branch tube will be brazed or welded.

8.1.1 Asymmetric Compression Beading

Figure 8.2 presents the fundamentals of the cost competitive tube branching process developed by Alves and Martins [1] that makes use of out-of-plane local buckling for joining tubes by means of asymmetric compression beads. Asymmetric compression beading works at room temperature and is accomplished by forcing one tube end towards the other (or the two tube ends towards one another) while leaving a gap opening in-between the dies that support and hold the tubes. The tube collapses at the gap opening creating the required asymmetric bead upon compression by the upper die.

On the contrary to axisymmetric beads, which are naturally formed by local buckling (during successive in-plane instability waves) in tubes subjected to axial loading between parallel flat dies, asymmetric beads require the development

Fig. 8.2 Asymmetric compression beading of thin-walled tubes. **a** Schematic representation of the process and **b** commercial S460MC carbon steel tube showing an asymmetric bead

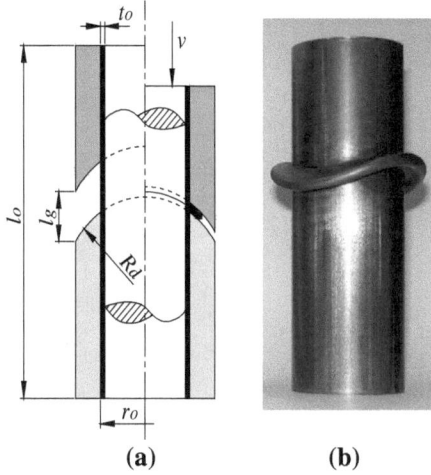

(a)   (b)

of out-of-plane instability waves between contoured dies [2]. A tool set-up for producing out-of-plane instability waves in tubes consists of two (upper and lower) contoured dies and an inner mandrel (Fig. 8.2a). The dies are dedicated to a specific reference radius r_0 of the tube. Their geometry, together with the initial gap opening l_g between them, is responsible for defining the shape and position of the beads.

The asymmetric compression bead shown in Fig. 8.2b was performed in a commercial S460MC carbon steel tube that was formed in the 'as-received' condition. The stress–strain curve of the S460MC tubes was determined by means of tensile and stack compression tests performed at room temperature on a universal testing machine with a cross-head speed equal to 100 mm/min (refer to Sect. 7.1),

$$\sigma = 616.4\varepsilon^{0.06} \ [MPa] \tag{8.1.1.1}$$

Numerical simulation of asymmetric compression beading involved a standard discretization procedure based on the utilization of approximately 7,500 eight-node hexahedral elements. In order to ensure the incompressibility requirements of the plastic deformation of metals, both complete and reduced Gauss point integration schemes were utilized. Dies and mandrel were discretized by means of contact-friction spatial linear triangular elements. The effects of strain rate and anisotropy on material flow behavior were neglected.

Convergence studies with varying arrangements of elements in the thickness direction showed that the utilization of three elements was adequate for modeling the distribution of the major field variables and for getting a proper evolution of the load–displacement curve.

The finite element predicted evolution of the out-of-plane instability wave for a test case performed with a mandrel inside the tube (Fig. 8.3) illustrates the key role played by the upper and lower contoured dies in establishing the final shape of the instability wave and the limits of its propagation path. The utilization of a

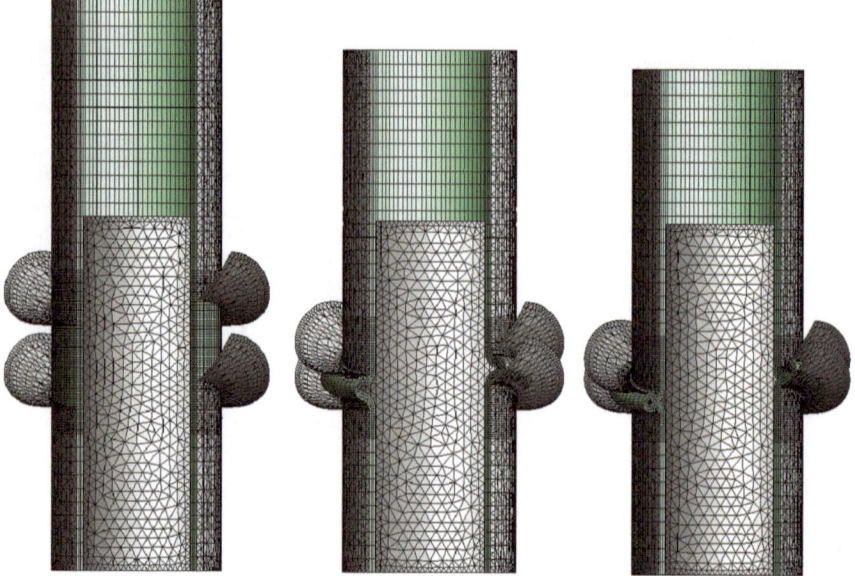

Fig. 8.3 Finite element predicted evolution of the out-of-plane instability wave for a test case performed with a mandrel

mandrel inside the tube not only avoids defects along the surface of the tube but also guarantees the dimension of the inner diameter (which in many applications is a critical dimension) to stay within tolerances. In practical terms, the use of a mandrel eliminates inward material flow and forces the asymmetric beads to develop exclusively outwardly.

Figure 8.4 shows the finite element predicted and experimental evolution of the load–displacement curve for the component shown in Fig. 8.2b. As seen, both evolutions compare well and allow distinguishing three different forming stages: (i) triggering the out-of-plane instability wave (labeled 'A' in Fig. 8.4), (ii) shaping the asymmetric compression bead from the out-of-plane instability wave (labeled 'B') and (iii) contacting of opposite sides of the asymmetric compression bead between the upper and lower contoured dies at the end of the stroke (labeled 'C').

In the first stage, the load increases steeply as the tube starts being axially compressed. A peak load of approximately 100 kN is obtained after which the load drops and the out-of-plane instability wave progressively begins to create the asymmetric compression bead in the free gap opening between the contoured dies.

The value of the peak load is similar to the critical instability load $P_{cr} \cong 93.5$ kN for the occurrence of axisymmetric local buckling (that is plotted as a horizontal dashed line) because there is almost no difference between in-plane and out-of-plane instability waves at the early stages of axial compression (refer to Fig. 8.3). The drop in load that is registered throughout the second stage (refer to 'B' in Fig. 8.4) is justified by the fact that forms of equilibrium resulting

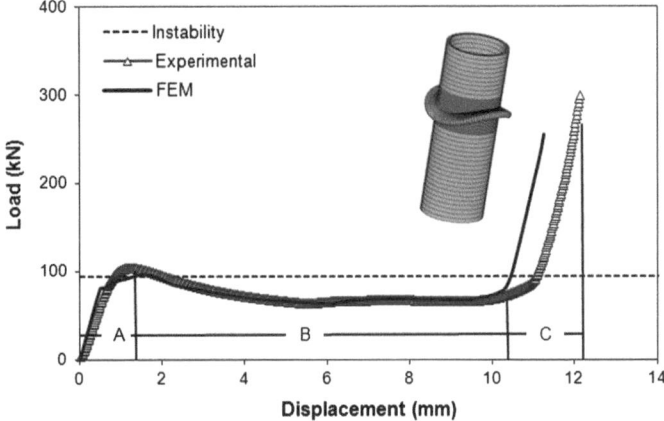

Fig. 8.4 Experimental and finite element predicted evolution of the load–displacement curve

from local buckling necessitate small values of the axial compressive load as the degree of instability increases.

The final sudden increase in the forming load during the third stage is triggered when the opposite sides of the compression bead get in contact between the two dies (refer to 'C' in Fig. 8.4).

8.1.2 Application

The utilization of two opposite asymmetric compression beads allows joining (locking) tubes by plastic deformation at room temperature and is the basis for the new tube branching process that was utilized for producing the tee fitting shown in Fig. 8.5.

8.2 Resistance Spot Welding

Resistance spot welding is a key technology in automotive assembly production, and it is by number the most used welding process. According to Zhu et al. [3], more than 200 sheet metal parts are spot welded together resulting in 4,000–7,000 spot welds of two and three sheet combinations in each car.

The development of new materials [such as e.g., advanced high strength steels (AHSS)] presents challenges to the resistance spot welding process when combined with other materials. These new steel types are often used in supporting parts of the car and in safety parts that are designed to absorb the impact of a crash. The parts are typically joined to considerably thinner and softer low-carbon sheet materials that act as the outer panels of the car. There is therefore an increasing trend

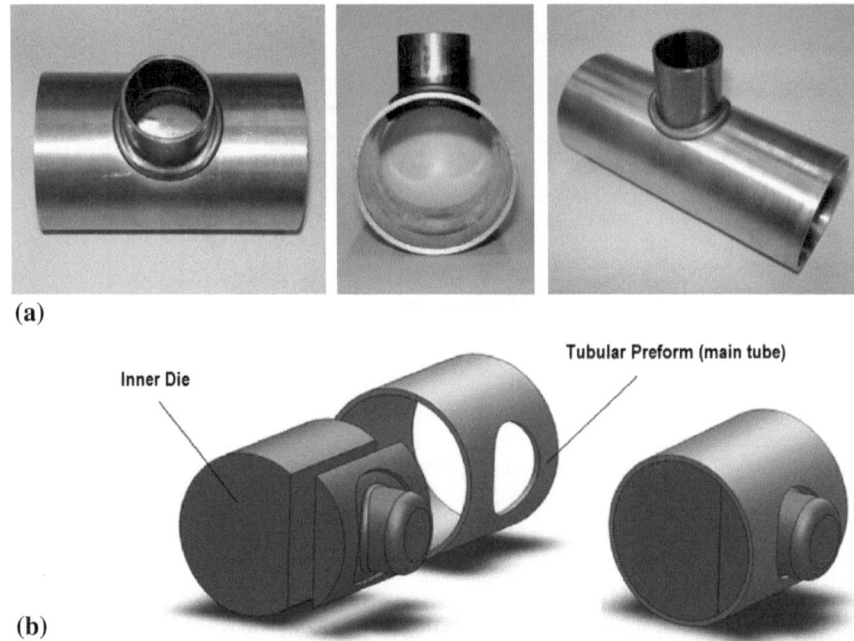

(a)

(b)

Fig. 8.5 Tube branching by means of asymmetric compression beading. **a** Typical tee fitting produced with the proposed joining process and **b** schematic representation of the inner sectioned die and preform of the main tube

of assembling three sheets by spot welding, which typically involves two thicker, high strength steels and one, thin mild steel as one of the outer sheets. This combination has attracted a lot of attention because of the difficulties in attaining a weld nugget at both interfaces as illustrated by Nielsen et al. [4].

The following three subsections deal with different challenges in resistance spot welding. An example consisting of three sheets as described above is dealt with in the first subsection by comparison of simulation and experiment. The second subsection elaborates on the same example by showing the effect of electrode misalignment, which is an important issue in production where the flexibility of the welding gun arms can result in a slight rotation of the electrodes. Another complication in industrial spot welding is the shunt effect between two consecutive spots. This is illustrated in the third subsection by spot welding a two sheet assembly.

8.2.1 Three Sheet Spot Welding

As already outlined above, spot welding of three sheets is the main challenge in automotive spot welding. Two thicker high strength steels and a thin low carbon steel is the typical combination. The specific combination chosen in the present

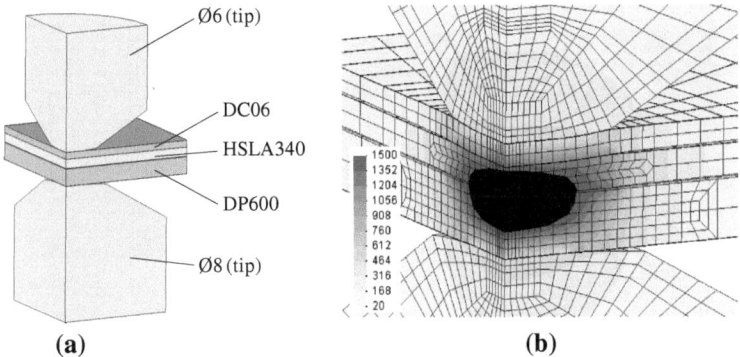

Fig. 8.6 Example of three sheet spot weld consisting of a thin 0.6 mm DC06 steel sheet, a 0.8 mm HSLA340 steel sheet and a 1.5 mm DP600 steel sheet welded between two type B electrodes with tip diameters Ø6 mm and Ø8 mm. **a** Quarter of the geometry showing material combination. **b** Detail showing the finite element predicted temperature field

example consists of a 1.5 mm DP600 dual phase steel (advanced high strength steel) as the bottom sheet, a 0.8 mm HSLA340 (high strength low alloy) steel in the middle and a 0.6 mm DC06 (low carbon deep drawing steel) as the top sheet. This combination is welded between two type B conical electrodes with tip diameter Ø8 mm towards the DP600 and tip diameter Ø6 mm towards the DC06 as illustrated in Fig. 8.6a with its finite element discretization and numerically predicted weld nugget in Fig. 8.6b. This example is reproduced from Nielsen et al. [4] with weld settings as follows. The weld force is constant 3.5 kN and the weld current is applied during 180 ms at 7.2 kA RMS through an AC weld machine with estimated conduction angle of 75 %.

As the DC06 sheet is considerably thinner than the DP600 sheet, the interface between the DC06 and the HSLA340 is located closer to the neighboring electrode than the interface between the DP600 and the HSLA340 is to its corresponding neighboring electrode. This results in larger heat conduction to the upper electrode and thus an asymmetric heat distribution. In the particular case (Fig. 8.6b), the heat input was too small to create a nugget that develops into the thinner sheet. On the other hand, if the heat input was too large, splash would be likely to occur between the two thicker sheets, leading to uncontrollable material removal, loss of strength, and excessive electrode wear. Compared to welding of two sheets, these restrictions result in a rather narrow window of applicable weld settings.

The simulated temperature distribution and weld nugget are compared to the corresponding experiment in Fig. 8.7a (cf. [4]). The overall weld nugget size is matching between the experiment and the simulation, and of specific interest in this case is that the finite element simulation reproduces the fact that the nugget does not develop into the thin sheet. This is in many cases a reason to reject the weld settings in order to achieve a weld nugget that covers both interfaces.

Due to the narrow window of appropriate weld settings (if any), innovative solutions have been developed to initiate the weld nugget in the interface towards

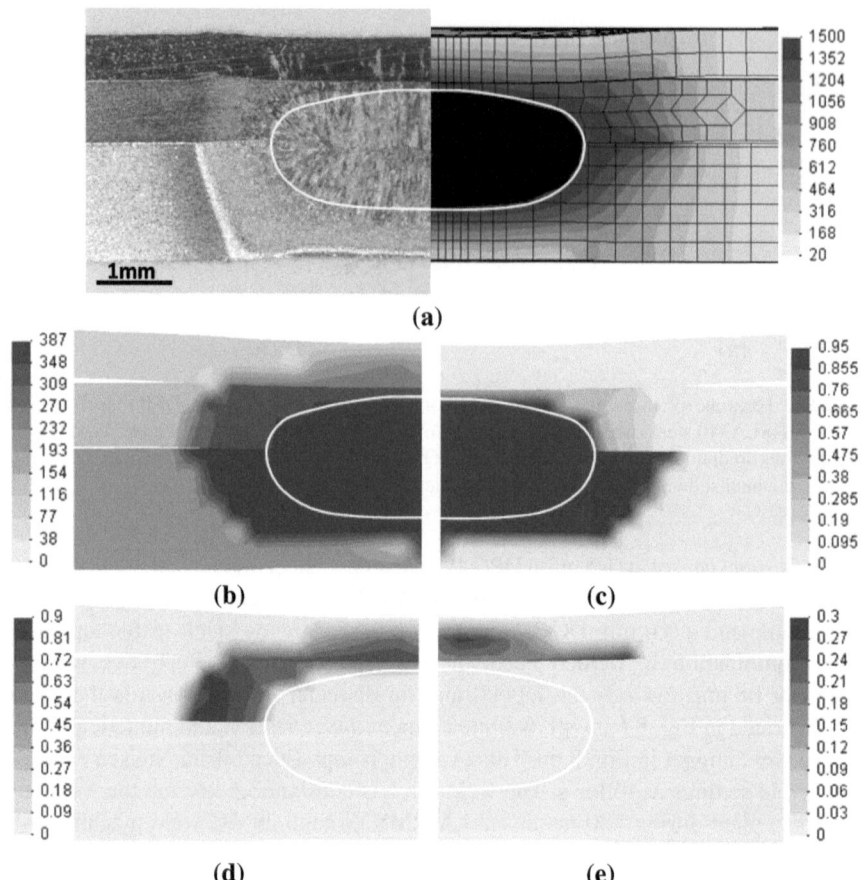

Fig. 8.7 Three sheet spot welding industrial test case. **a** Comparison between experimental and simulated maximum temperature distribution with indication of simulated weld nugget mirrored onto the experiment. **b** Simulated hardness distribution in Vickers. **c** Simulated martensite distribution. **d** Simulated bainite distribution. **e** Simulated pearlite distribution

the thin sheet as noted by Nielsen et al. [4], who at the same time proved that plug failure mode (the desired failure type in tensile-shear testing) can be achieved without melting into the thin sheet. The strength in the interface towards the thin sheet is in those cases achieved by solid state bonding facilitated by heat and plastic deformation.

Having the temperature history simulated including maximum temperatures and cooling rates as well as knowing the compositions of the base materials [4], it is possible to calculate the resulting hardness distribution (Fig. 8.7b) and microstructure distributions. The individual fractions of selected phases of the microstructure are shown in terms of martensite (Fig. 8.7c), bainite (Fig. 8.7d) and pearlite (Fig. 8.7e) and demonstrate the potential of multi-object simulation by means of the electro-thermo-mechanical coupled finite element flow formulation to predict the metallurgical behavior of materials.

Following time–temperature–transformation (TTT) diagrams for the specific steels, the fractions of the different phases are found by comparison with critical cooling rates. Typically, and also in Fig. 8.7c, the center of the nugget consists mainly (here 95 %) of martensite due to the prior full transformation into austenite followed by rapid cooling. Outside the nugget, the material may form bainite and pearlite depending on the initial composition and the actual cooling rates (Fig. 8.7d, e) or more martensite as in the DP600 steel.

The estimation of the quantities inside the nugget is complicated by the presence of more than one material. The contribution of each material to the combined microstructure and hardness distribution is evaluated by volume weighting assuming that the material inside the nugget is fully mixed in its molten stage.

The hardness is evaluated by the model by Blondeau et al. [5] based on the actual cooling rate and the carbon equivalent. More details of the evaluation of microstructure and hardness distributions can be found in the work by Pedersen et al. [6], who also compare experimental and simulated results.

8.2.2 Electrode Misalignment

The above industrial case with three sheets will now be analyzed under the assumption of electrode misalignment, which is relevant to assembling in a production line. A potential source of electrode misalignment is the flexibility of the welding machine arms for positioning the electrodes. These arms are necessary in order to reach the locations of the spots on larger panels. Rotation of the electrodes can occur as illustrated in Fig. 8.8 when applying the electrode force through these arms.

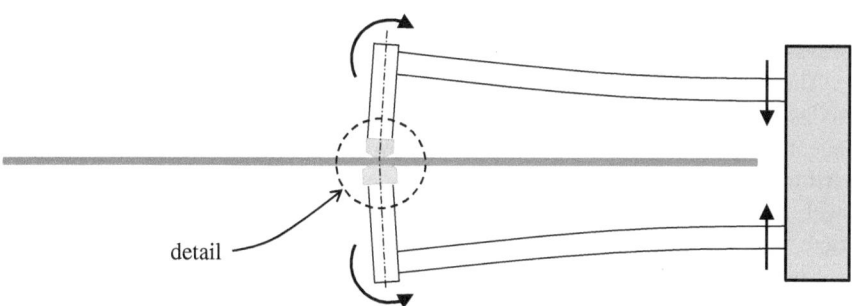

Fig. 8.8 Electrode misalignment due to rotation caused by flexibility of the arms of a welding gun typically applied in production in order to reach the location of the spots. The detail enclosing the electrodes is enlarged in Fig. 8.9a

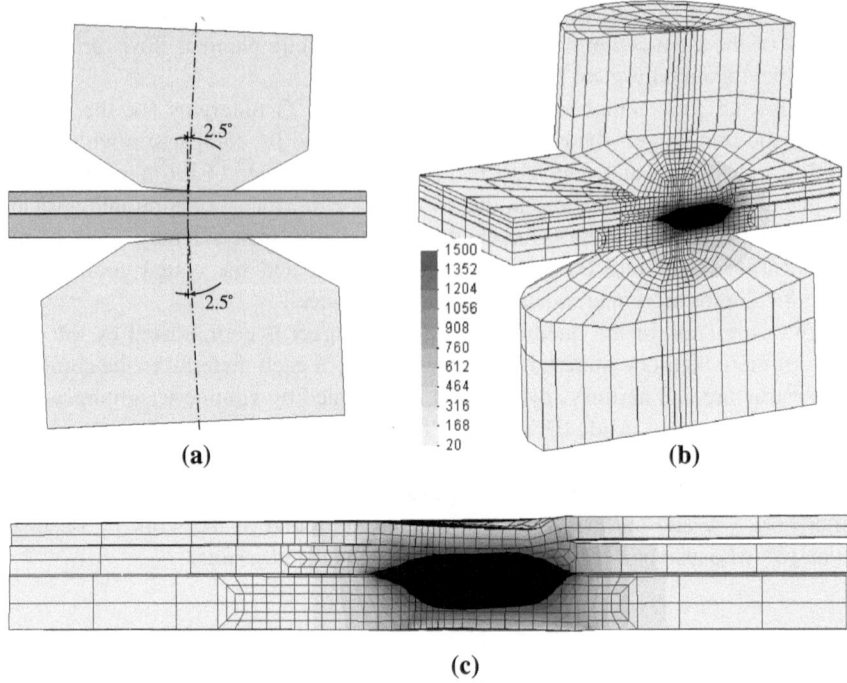

Fig. 8.9 Electrode misalignment in the three sheet spot welding case of Sect. 8.2.1. **a** Electrode misalignment by rotation of each electrode by 2.5°. **b, c** Simulated peak temperature distribution with indication of asymmetric weld nugget and visible excessive electrode indentation of the upper electrode into the upper thin sheet (**c**)

A situation like the one illustrated in Fig. 8.8 is simulated with the same sheet setup and weld settings as in the above case. The electrode rotation is assumed to be 2.5° for each of the electrodes as illustrated in Fig. 8.9a showing the detail of Fig. 8.8. The finite element model including the electrode misalignment consists of 5,542 elements and Fig. 8.9b shows the simulated peak temperature field and indication of the weld nugget. The weld nugget can be compared to Fig. 8.7a because all other parameters than the electrode misalignment are identical.

The resulting weld nugget is clearly asymmetric as a result of the angled electrodes which are initially only touching the sheets on the outer edge until a certain indentation has developed. Figure 8.9c shows a close up of the weld nugget as well as a clear angled indentation of the upper electrode into the thin low carbon steel sheet. By comparison to the symmetric weld in Fig. 8.7a, the indentation is more severe in case of electrode misalignment due to the small initial contact area.

The larger and localized indentation causes the thin sheet to lift more (right side in Fig. 8.9c), and complicates the overall assembly process because distortion can create relative movement of the sheets to a degree that makes the sheets off position at the location of following spots. This is already an issue under ideal conditions that need to be taken care of in the planning of the sequence of the welds. This procedure is further complicated by the additional distortion due to eventual misaligned electrodes.

The simulation also shows that the gap between the two high strength steels is increased by the introduction of angled electrodes, while at the same time the nugget forms towards the gap opening. This increases risk of splash significantly, which would lead to uncontrolled joining conditions.

From the discussion in Sect. 8.2.1, it is clear that the chosen weld settings are too low to form a weld into the thin upper sheet, and that the weld settings should be increased (i.e., increased current/weld time or lowered electrode force). However, due to the lowered splash limit by the electrode misalignment, it might not be possible to increase the weld settings in this case. Formation of a weld nugget into the thin sheet may therefore be impossible in case of electrode misalignment (leaving out of account innovative solutions to initiate the weld nugget at the interface towards the thin sheet).

8.2.3 Shunt Effect

Shunt effect is taken as another complication occurring in industrial joining with multiple spot welds. The effect is considered in a case with two sheets welded between two type B electrodes (cone shaped as in Fig. 8.6a) with tip diameter Ø6 mm. The two sheets are chosen to be different steels with different thicknesses. The bottom sheet is a 1.2 mm DP600 steel and the upper sheet is a 0.7 mm DC06 steel.

The squeeze time is simulated as 40 ms to reach the constant welding force 2.5 kN. The AC welding current is kept constant at 8 kA RMS for 160 ms, such that the welding current of the first spot is ending at time 200 ms (temperature field shown in Fig. 8.10). The electrode force is kept during a hold time of 80 ms finishing the first weld at time 280 ms. Hereafter follows 3 s where the electrodes are repositioned to the location of the second spot (temperature fields at selected instants of time (370, 1,310, 2,230, and 3,190 ms) during the repositioning are shown in Fig. 8.10).

The location of the center of the second spot is 12 mm away from the center of the first spot. This distance corresponds to two electrode tip diameters, which is closer than the recommended minimum distance between spots. This is chosen in order to magnify the shunt effect to support the presentation. After moving the electrode, the squeeze time is again initiated and the same weld schedule is applied as for the first weld, implying that also the weld current is kept at the same level. The shunt effect is therefore not compensated by an increased current and a comparison of the weld nuggets will show the effect of shunting. The weld current of the second weld ends at time 3,480 ms (temperature field shown in Fig. 8.10).

The above referred temperature fields and instants of time are collected in Fig. 8.10 to give an overall representation of the shunt effect. The upper left temperature field shows the ending of the first weld current. Following the arrows, the

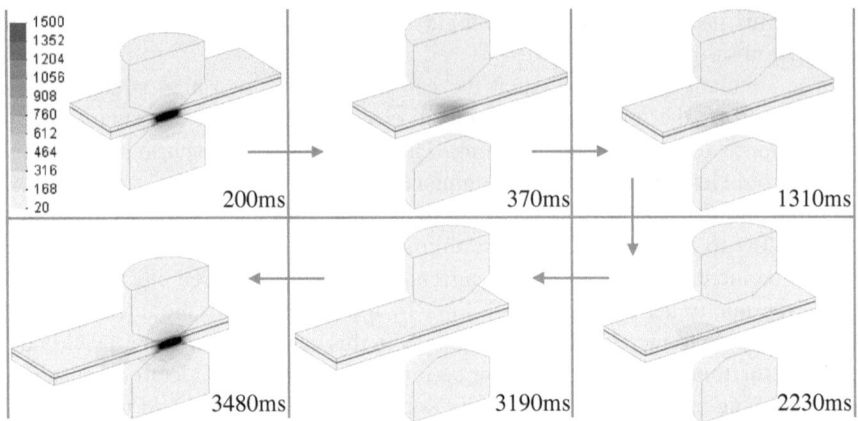

Fig. 8.10 Shunt effect between two consecutive spot welds illustrated by the temperature field at different instants of time

following four temperature fields illustrate the temperature evolution during the movement of the electrodes to the location of the second spot. The first weld cools while the surrounding sheet material is moderately heated due to heat conduction. The last temperature field shown in the figure corresponds to the ending of the second weld current. At this stage the second weld nugget has formed, but also the temperature in the first spot has risen as seen by a comparison between the two last instants of time. This is due to electrical heating caused by the shunting current flowing through the first spot while welding the second spot.

The shunting current is shown in Fig. 8.11a by the current density at the peak current of the third half cycle. While the majority of the current flows through the sheet interface at the location of the second spot, it is seen that a considerable amount of current flows through the first spot because of the absence of an interface after welding. At the location of the first weld, the current density is seen to be higher where the sheets start to separate towards the second spot due to the singularity. The amount of shunting current varies during the welding time of the second spot. The contact resistance between the sheets is larger at low temperatures indicating a larger shunting current in the beginning, but on the contrary the bulk resistivity increases with temperature, which indicates a larger shunting at the later stages because the material between the spots as well as the first spot remain at moderate temperature.

The peak temperature distribution achieved during the second weld is shown in Fig. 8.11b, where it is shown that the temperature in the first spot raises to 330 °C during the second spot welding. As a result of the shunting current and the temperature increase in the first spot, less heat is dissipated in the second weld compared to the first one. This is also directly readable from the resulting weld nuggets shown in Fig. 8.11c where the overall peak temperature distribution of the entire welding process is shown. The nugget sizes measured at the interface between the sheets are 4.84 mm in the first spot and 4.18 mm in the second spot, which is a decrease of 14 %.

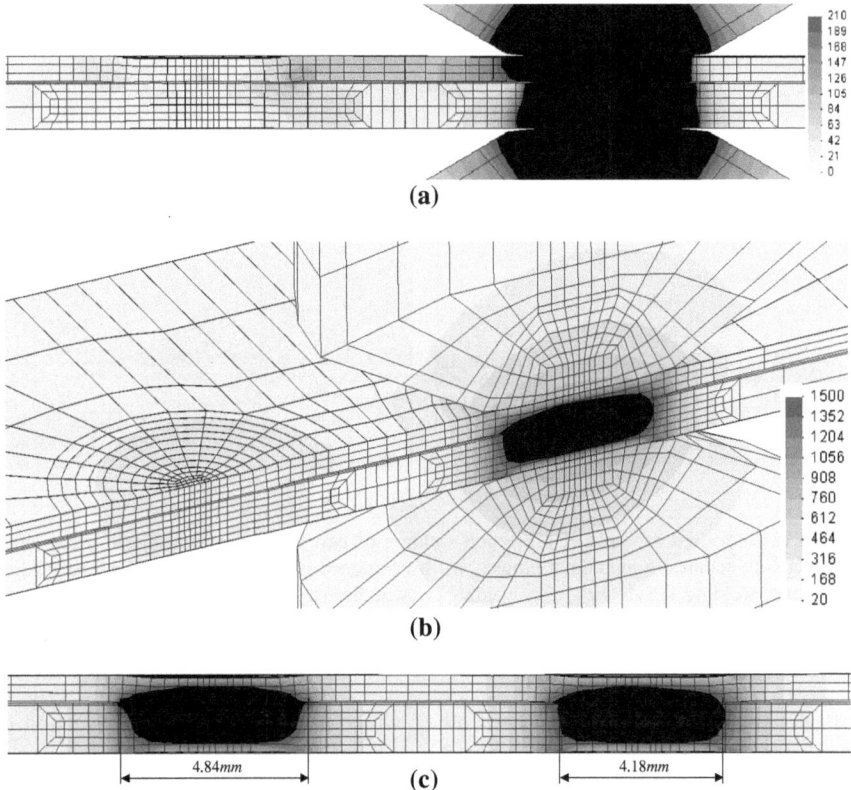

Fig. 8.11 Detail of the finite element predicted simulation of the shunt effect between two consecutive spot welds of two sheets. **a** Current density after 5/4 cycles of the second weld current on a 0–210 A/mm² scale out of maximum 1683 A/mm². **b** Peak temperatures reached during the second weld showing the weld nugget achieved in the second weld while reaching 330° C in the first spot. **c** Comparison of the two weld nuggets by overall peak temperatures (first spot to the *left* and second spot to the *right*)

In a production line where the shunt effect will play a role due to the necessity of having spot welds located close to each other, the current can be increased to compensate for the heat dissipated in the neighboring spot(s).

8.3 Projection Welding by Longitudinal Embossment

Projection welding is another important variant of the resistance welding processes, where the current is concentrated to the weld region by a projection (natural or fabricated) instead of being concentrated through the electrodes as in the above spot welding examples. Figure 8.12 shows an application of projection welding

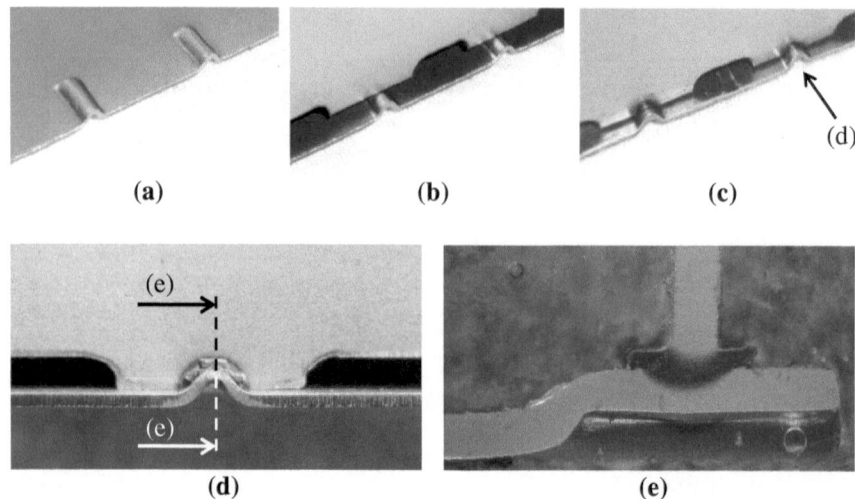

(a)   **(b)**   **(c)**

(d)   **(e)**

Fig. 8.12 Industrial example of projection welding of two sheets perpendicular to each other. **a** Sheet with stamped longitudinal projections. **b** Positioned perpendicular sheets before welding and **c** after welding. **d** Side view after welding (view indicated in (**c**)). **e** Cross-section as indicated in (**d**)

of two sheets perpendicular to each other, which is relevant to e.g., fabrication of housings and containers that are not required to be water or air tight and to the addition of perpendicular stiffeners to sheet panels.

The presented example is an industrial case provided by a Japanese company. In order to facilitate joining of two sheets perpendicular to each other by projection welding, one of the sheets is embossed as shown in Fig. 8.12a. When the other sheet is positioned as shown in Fig. 8.12b, the longitudinal embossments ensure local contacts between the two sheets. Resistance projection welding is carried out under constant weld force and DC current resulting in the joint shown in Fig. 8.12c. A close up of one of the projection welds is shown in Fig. 8.12d and a cross-section is shown in Fig. 8.12e.

The two sheets are 0.8 mm thick high strength low alloy steel sheets (grade similar to HSLA340). The welding parameters are as follows: 700 N weld force, 3.5 kA DC weld current and 30 ms weld time.

Figure 8.13a shows a finite element discretization of one of the projection welds by 5,630 hexahedral elements for simulation of the above industrial case. The simulation utilizes a natural symmetry plane along the longitudinal projection (that is the cutting plane utilized to show the cross-section in Fig. 8.12e). An additional symmetry plane is assumed in the simulation to reduce the model size. It is introduced in the center of the vertical sheet in Fig. 8.12e, such that the final model utilizing both symmetry planes is as shown in Fig. 8.13a. The round part of the sheet is only to make structured meshing easier of the round end of the projection. It does not influence the simulation due to the distance from the weld.

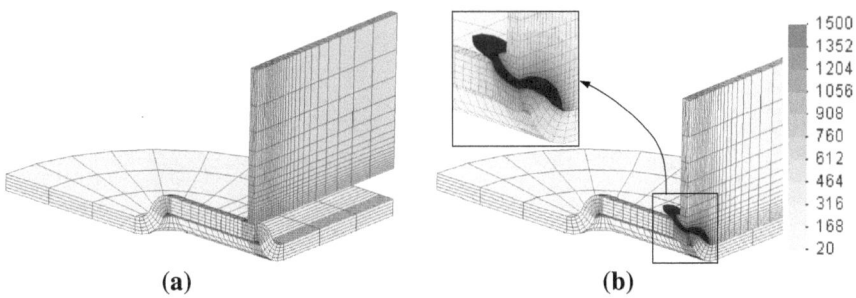

Fig. 8.13 Projection welding of two perpendicular sheets. **a** Initial finite element mesh and **b** predicted peak temperature field at the end of the welding showing molten volume squeezed out between the two sheets

The second symmetry plane is justified as follows with reference to Fig. 8.12e. The differences on each side of the vertical sheet in terms of the electrical and thermal fields are considered negligible. This is in light of the short process time (weld time is 30 ms) and considering the distance to the end of the embossment on one side and to the end of the bottom sheet on the other side. As regards the mechanical aspects of the assumed symmetry plane, the geometry after welding (Fig. 8.12e) is symmetric around the vertical sheet. The longitudinal embossment does not bend towards the free end and therefore the free end can be omitted from the simulation. The side including the rounded end of the embossment is included to prevent the embossment from flattening, and the mirroring of that does not affect the overall deformation.

The simulated weld is shown in Fig. 8.13b with the peak temperature distribution shown. In the interface of the two sheets, the material melts and squeezes out as in the real case (compare detail in Fig. 8.13b to the cross-section in Fig. 8.12e) while the upper sheet closes towards the bottom sheet (compare Fig. 8.13b to the side-view in Fig. 8.12d).

A detailed comparison of the real example (Fig. 8.12) and the simulated projection weld (Fig. 8.13) is presented in Fig. 8.14 in the cross-section similar to Fig. 8.12e. The comparison covers the final geometry as well as the temperature history. As regards the geometry, the main difference is the shape of the metal that is squeezed out between the two sheets in a molten or mushy state. The exact shape might be of less importance compared to the volume squeezed out and the formed contact area during welding as it relates to the heat development. In the specific example, more elements would be required in the volume that is squeezed out if the details of the squeeze out are of importance.

The heat development and the heat balance were of more importance when doing the presented simulation in collaboration with the company. The comparison of the temperatures can be facilitated by the simulated temperature field and the resulting microstructure of the real case. The simulated temperatures shown in the simulation are the peak temperatures achieved during the simulation, and these can be related to the changes in microstructure. The selected isothermal lines in

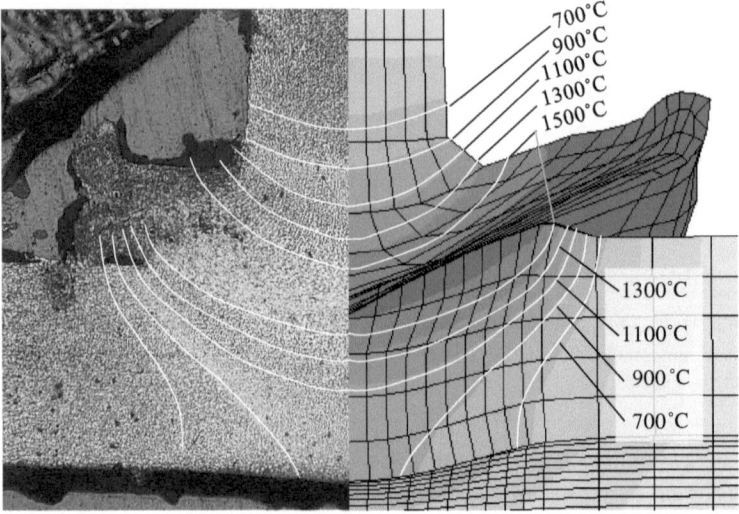

Fig. 8.14 Comparison of cross-section of the real component and simulated peak temperature distribution in the cross-section view similar to Fig. 8.12e. The simulated peak temperature is shown on a 20–2,000 °C scale with selected isothermal lines. These lines are mirrored onto the actual cross-section

the simulated temperature field are mirrored onto the cross-section of the real case, revealing that the temperature gradients are simulated correctly as the isothermal lines of the simulated temperature field match the shape of the border lines between the different microstructures.

8.4 Welding of Bellow to Disc by Natural Projection

This section presents an industrial resistance welding example from a Danish company. The specific welding case is part of the production of thermostat valves for radiators. Inside the thermostat valve is a bellow that expands or contracts due to temperature changes and thereby opens or closes the valve controlling the heating of the radiator. A few steps of this production are illustrated in Fig. 8.15.

A tin-bronze bellow tube with a conical collar (2) is resistance welded to a steel ring (3) between electrodes (1) and (4) as schematically shown in Fig. 8.15a by its setup. A result of this welding process is the joined bellow tube and steel ring shown in Fig. 8.15b (upside down compared to Fig. 8.15a). The bellow is hereafter formed as shown in Fig. 8.15c before it is mounted in a container as depicted in Fig. 8.15d (turned back to the same orientation as Fig. 8.15a). The joint between

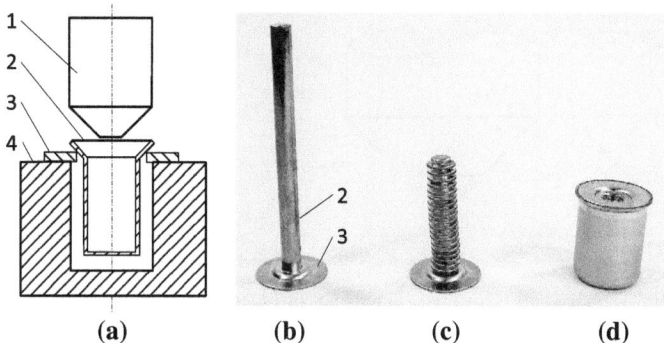

Fig. 8.15 Selected process steps utilized in the production of thermostat valves. **a** Resistance projection welding of bellow tube to steel ring. **b** Joined bellow tube and steel ring. **c** Formed bellow. **d** Mounting in container by resistance projection welding of steel ring to container

the steel ring and the container is also accomplished by resistance projection welding. This section focuses on the resistance projection welding of the bellow tube to the steel ring (Fig. 8.15a, b).

The Ø8 mm bellow tube is produced in tin-bronze CuSn6 (W.Nr. 2.1020) with a wall thickness of 0.14 mm. A 90° conical collar is formed prior to welding such that the contact to the steel ring forms a natural projection. The 1 mm thick mild steel (W.Nr. 1.0338) ring has outer diameter Ø29 mm and hole diameter Ø8.3 mm and it is coated with a 2–6 µm thick layer of electroless deposited Ni–P alloy (8–12 % P) to facilitate welding. The upper 90° conical electrode is a standard copper alloy for resistance welding, CuCr1Zr, A2/2 after ISO 5182:1991.

This welding case was analyzed by Rasmussen [7] and Bay et al. [8] with focus on electrode wear and the influences on the weld quality. The joint is tested for leakage in the production by an applied pressure. Very few (of the order of per thousand) defects are observed when welding up to 40,000 pieces, but the defect rate increases with electrode wear. The influence of electrode wear is therefore analyzed and presented in the following by new finite element simulations.

The electrode geometry changes significantly due to electrode wear as illustrated by a new and a worn electrode in Fig. 8.16a, b. A cross-section of a worn electrode after 580,000 welds is shown in Fig. 8.16c showing severe change in electrode geometry from the original conical shape. This number of welds is well beyond the normal tool life and is made for the analysis such that clear effects are noticed.

The differences in the resulting welds are analyzed by a combined metallographic study of selected cross-sections and a numerical study based on the electro-thermo-mechanical finite element flow formulation that has been presented throughout the book. The numerical study is based on the axisymmetric finite element models shown in Fig. 8.17. The setup including a new electrode is shown in Fig. 8.17a with a close-up of the simulated deformation and temperature in the end of the weld time in Fig. 8.17b. The model including a worn electrode is shown in

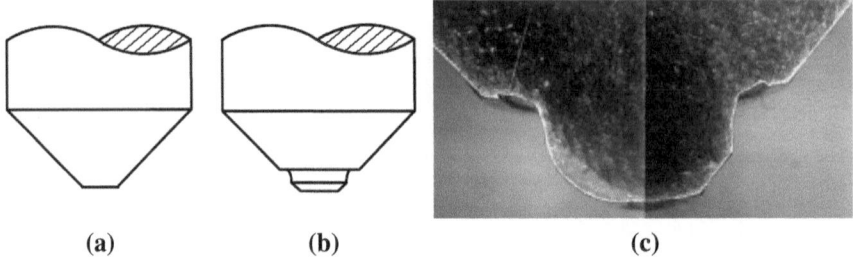

Fig. 8.16 Illustration of electrode wear. **a** New electrode. **b** Model of a worn electrode. **c** Cross-section of a real worn electrode after 580,000 welds, which is well beyond the normal tool life

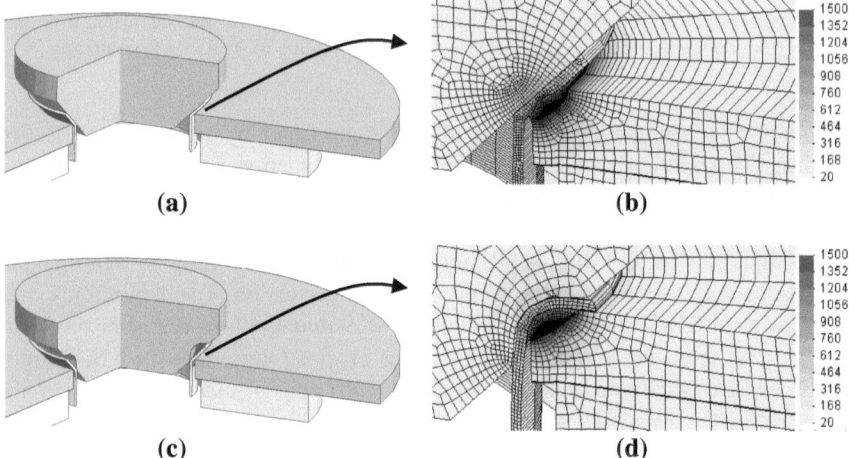

Fig. 8.17 Simulated projection welding of bellow tube to steel ring with new and worn electrode. **a** Model including new electrode with detail shown in **b** including the simulated deformation and final temperature field. **c** Model including worn electrode with detail shown in **d** including simulated deformation and final temperature field

Fig. 8.17c with the shape of the electrode equal to the worn electrode shown by its cross-section in Fig. 8.16c. Figure 8.17d shows the simulated deformation and final temperature as a result of welding with the worn electrode.

The simulated temperature fields of Fig. 8.17b and Fig. 8.17d are compared with the cross-sections of the real welds in Fig. 8.18 contributing to the overall analysis. Deformation and microstructure show clear differences between the welds stemming from a new electrode and the worn electrode.

After welding with a new electrode, an investigation of the microstructure (Fig. 8.18a) shows that the steel adjacent to the weld interface has a very coarse, ferritic grain structure, which is observed as a thin, bright zone. Apparently, the steel in this zone has been heated close to, but not above, 900 °C leading to grain growth in the ferrite. This is confirmed by the corresponding numerical simulation

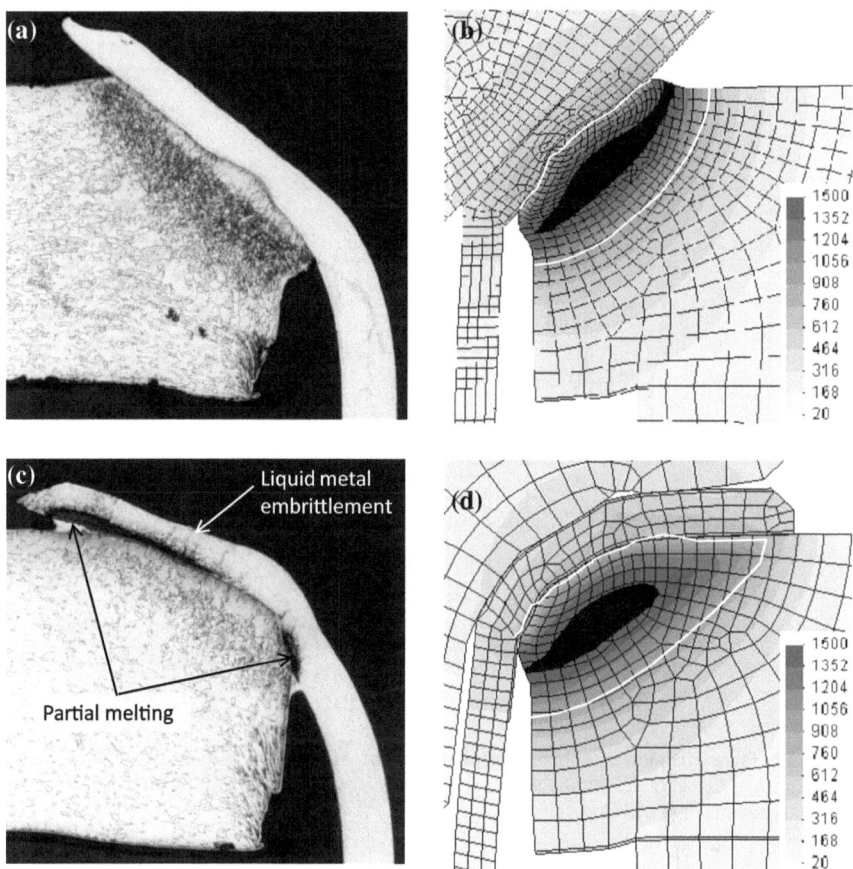

Fig. 8.18 Cross-sections of welded bellow tube to steel ring. **a** Weld performed with a new electrode. **b** Simulation with new electrode. **c** Weld performed by worn electrode and indication of partial melting and liquid metal embrittlement in the bellow tube. **d** Simulation with worn electrode. Figures **b** and **d** show the final temperature field together with contour lines corresponding to the 900 °C isothermal line of the peak temperature field

(Fig. 8.18b), where the white isothermal line corresponds to 900 °C peak temperature. This isothermal line is seen to leave a small gap to the interface confirming the above hypothesis.

Below this zone, the microstructure appears to be dark indicating that the peak temperature in this part of the steel has been raised to above 900 °C causing a phase transformation to fine-grained austenite. During the subsequent rapid cooling this austenite transformed to very fine-grained ferrite, which appears dark on the micrograph. The microstructure of Fig. 8.18a indicates that the highest temperature during the welding process was reached inside the steel at a certain distance from the weld interface, and not at the interface itself. This is due to the large

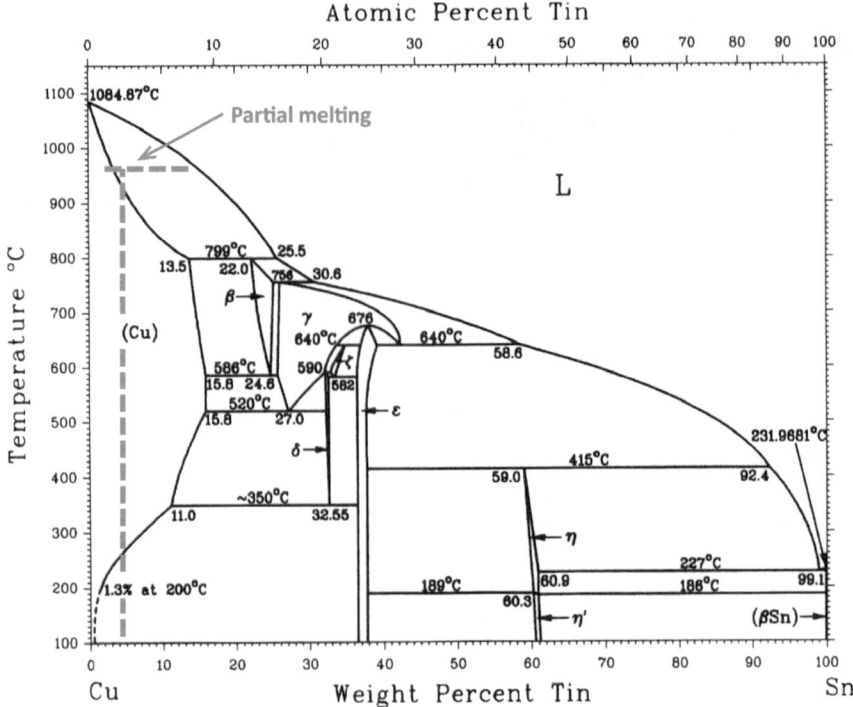

Fig. 8.19 Phase diagram of copper (Cu) and tin (Sn) with the actual tin-bronze alloy marked by the *dashed line* at 6 weight-% tin. The specific alloy experiences partial melting in the region above approximately 900 °C

difference in electric resistivity of the tin-bronze bellow tube and the steel ring. This effect is also seen in the simulation (Fig. 8.18b). No phase transformations are observed in the tin-bronze bellow tube, which is thus kept in a good condition.

Welding with the heavily worn electrode results in a microstructure (Fig. 8.18c), which as regards the steel contains the same microstructural elements as those seen in Fig. 8.18a. Due to the poor contact between the upper electrode and the tin-bronze tube in the first phase of welding, the tube experiences higher temperatures than with a new electrode. The dark areas in the tin-bronze represent areas of partial melting and hot cracking. This occurs for the bellow tube material when the temperature is above approximately 900 °C, cf. the Cu–Sn phase diagram provided in Fig. 8.19 [9], where the actual tin-bronze alloy (CuSn6) is marked by the dashed line.

The simulated weld by the worn electrode (Fig. 8.18d) also reveals peak temperatures above 900 °C in the tin-bronze by the white isothermal line. The real weld is seen to have experienced heavier partial melting than the simulation shows. This can stem from asymmetric wear of the electrode, which will result in further localization of the heat along the circumference, whereas the axisymmetric simulation distributes the heat evenly along the circumference.

Besides the partial melting, liquid metal embrittlement is noticed in Fig. 8.18c. This is caused by penetration of melted Ni–P coating into the grain boundaries of the tin-bronze. These phase transformations are explained by the elevated temperatures reached when welding with a worn electrode.

8.5 Micro Joining of Fork and Wire

An industrial case from the electronics industry is provided by means of a collaborative work with a German company. The application is micro joining of a fork to a wire as shown in Fig. 8.20 in its configuration before joining. The wire (1) is pure copper of diameter $\varnothing 0.73$ mm coated by a 5 μm thick polyimide plastic (2). It is joined to an alloyed copper fork (3) between two tungsten electrodes (4 and 5). The tungsten electrodes close the fork legs around the wire by an applied force to form the joint. A current is simultaneously applied for two reasons. The resistance heating caused by the current facilitates the closing of the fork around the wire due to softening of the material. At the same time, the induced temperature melts the polymer coating locally on the wire to create electrical connection between the wire and the fork, which is required for the use of the component while the polymer keeps the remaining wire isolated.

Although no weld is created (there is no melting except for the coating), the joining is facilitated by the principles of resistance welding, and the multi-object numerical simulation based on the electro-thermo-mechanical coupled finite element flow formulation appears very effective for performing the analysis of the process. Two natural symmetry planes are utilized to reduce the finite element model, such that the finite element mesh in Fig. 8.21a consisting of 4,856 elements represents the overall geometry. The process conditions shown in Fig. 8.21b are applied in the simulation. The applied force is built up to a level of 120 N and kept constant until and during the first current pulse. The first current pulse has an up-slope time of 80 ms reaching 0.75 kA DC, which is kept constant for additionally 80 ms. The force is raised to 150 N before the second current pulse, which is applied as a constant current of 1.2 kA DC during 50 ms.

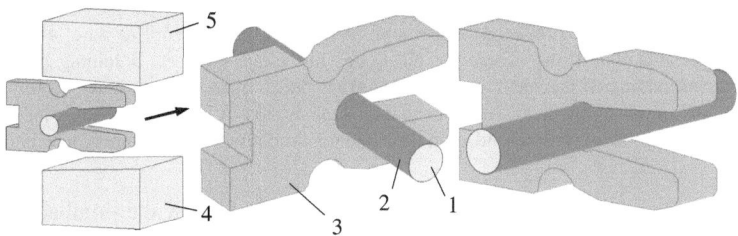

Fig. 8.20 Initial configuration of fork to wire joining process of copper wire *1* coated by polyimide plastic *2* to a copper alloy fork *3* between two tungsten electrodes (*4* and *5*)

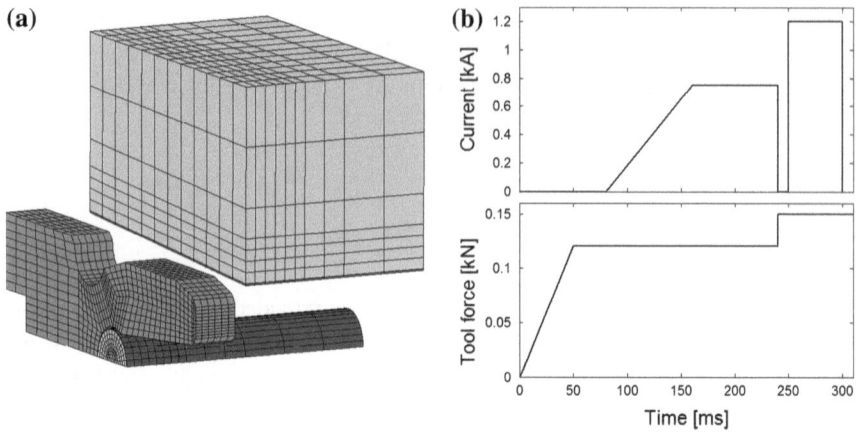

Fig. 8.21 Simulation of micro joining of fork to wire. **a** Initial mesh by utilization of two natural symmetry planes. **b** Applied current and force as function of process time

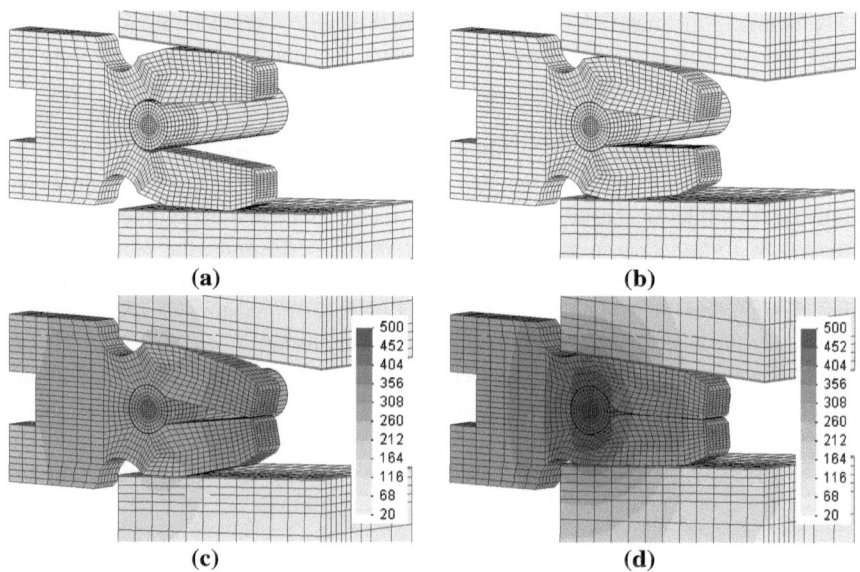

Fig. 8.22 Finite element predicted temperature in micro joining of fork to wire. **a** Start of the joining process at time 0 ms, where the electrodes just touch the fork. **b** Joining process after 80 ms corresponding to the onset of the first current pulse. **c** End of first current pulse at time 250 ms. Maximum temperature reached at this stage is 334 °C. **d** Completion of the joining process including the two current pulses at time 300 ms. Maximum temperature reached is 507 °C

The effect of the two applied pulses is shown in Fig. 8.22 by the simulated process. Figure 8.22a shows the moment where the tungsten electrodes just touch the legs of the fork. This corresponds to time 0 ms in Fig. 8.21b where the force is applied. After 80 ms, the applied force has been kept constant for 30 ms and the first current pulse

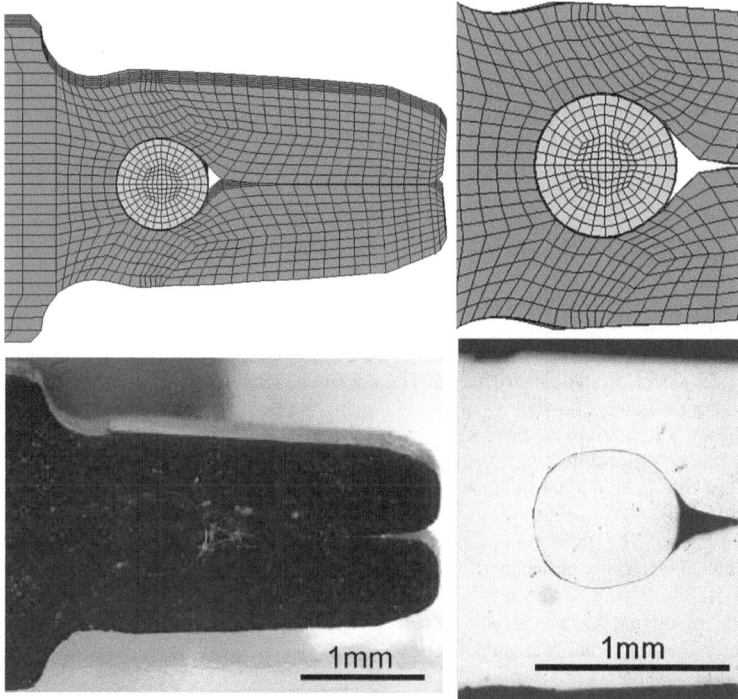

Fig. 8.23 Comparison of simulation (*upper*) and the real component (*lower*) in terms of the final geometry of the joined wire and fork

is ready to be applied. At this stage (Fig. 8.22b) the deformation of the fork is enough to close the initial gap towards the wire such that a sound contact is setup before applying the current. At the end of the first current pulse, the tips of the fork legs are closed (Fig. 8.22c). This deformation happens under the same applied force due to softening of the material. The temperature field after the first current pulse is shown in Fig. 8.22c with a maximum reached temperature 334 °C. In order to perform the final closing of the fork, the second pulse is applied while at the same time increasing the applied load. This results in the final geometry shown in Fig. 8.22d. The figure also shows the temperature field with a maximum reached temperature 507 °C, which is sufficient to melt the polymer coating to create electrical contact between the fork and the wire of importance to the final component.

The final geometry is compared to the real component in Fig. 8.23. The left figures compare the overall deformation showing that both the simulation and the real joint result in closing of the fork to a degree where the fork legs touch each other along the majority of their length. A detailed view of the region near the wire is shown in the right figures, where it is seen that the fork is closed around the wire with almost no deformation of the wire, which has part of its stiffness from the wire outside the contact area to the fork. The right figures also show that the amount of closing of the fork is simulated correctly near the wire.

References

1. Alves LM, Martins PAF (2012) Tube branching by asymmetric compression beading. J Mater Process Technol 212:1200–1208
2. Gouveia BPP, Alves ML, Rosa PAR, Martins PAF (2006) Compression beading and nosing of thin-walled tubes using a die: experimental and theoretical investigation. Int J Mech Mater Des 3:7–16
3. Zhu W-F, Lin Z, Lai X-M, Luo A-H (2006) Numerical analysis of projection welding on auto-body sheet metal using a coupled finite element method. Int J Adv Manuf Technol 28(1–2):45–52
4. Nielsen CV, Friis KS, Zhang W, Bay N (2011) Three-sheet spot welding of advanced high-strength steels. Weld J 90(2S):32s–40s
5. Blondeau R, Maynier P, Dollet J (1973) Prediction of the hardness and strength of plain and low-alloy steels from their structure and composition. Memoires Scientifiques de la Revue de Metallurgie 70(12):883–892
6. Pedersen KR, Harthøj A, Friis KL, Bay N, Somers MAJ, Zhang W (2008) Microstructure and hardness distribution of resistance welded advanced high strength steels. In: Proceedings of the 5th international seminar on advances in resistance welding, Toronto, Canada, pp 134–146
7. Rasmussen MH (2000) Kvalitetssikring af pressvejseprocessen (in Danish). Ph D thesis, Technical University of Denmark
8. Bay N, Zhang W, Rasmussen MH, Thorsen KA (2003) Resistance welding: Numerical modelling of thermomechanical and metallurgical conditions. DMS Winter Annual Meeting, Danish Metallurgical Society, pp 87–100
9. ASM International (1992) Alloy phase diagrams. ASM Handbook 3:2167–2182

Appendix A

The FORTRAN source code including OpenMP instructions for the parallel skyline solver is listed as follows:

```fortran
      subroutine skyline_gauss_omp (skmatx,fmatx,&
                                    maxa,nthreads,ntotv)

      use omp_lib

      implicit none

!     --------------------------------------------------
!     This subroutine solves a regular system:
!         skmatx*x=fmatx
!     where
!     skmatx is the skyline vector of the system
!            matrix,
!     fmatx  is the right hand side vector (in)
!            and later the vector of unknowns (out),
!     x      is the vector of unknowns outputted
!            through fmatx
!     The index vector pointing to the diagonal
!     positions in skmatx is maxa.
!     The number of degrees of freedom is ntotv.
!     Number of threads to use during solving is
!     nthreads.
!     Method: Gaussian elimination with column
!             reduction.
!
!     This skyline solver was originally provided
!     in sequential form by
!         J.E.Akin, Finite Elements for Analysis
!         and Design, Academic Press, London, 1993.
!     It is parallelized in the present work by
!         C.V.Nielsen and P.A.F.Martins
!     --------------------------------------------------
```

```
      integer i,id,ie,ie0,ie0old,ih1,ih2,ihesitate,&
              iloop,iquit,ir,is,ithread,iwait,j,jd,&
              jh,jmax,jr,k,k0,k00,kmax,nthreads,ntotv
      integer maxa(*)
      double precision d,fmatx(*),skmatx(*)

!         Set number of threads
      call omp_set_num_threads (nthreads)
!         Get (actual) number of threads
      nthreads=omp_get_max_threads()

!         Initializations
      jmax=1
      kmax=1

!$OMP parallel default (none) &
!$OMP private (d,i,id,ie,ie0,ie0old,ih1,ih2,ihesitate,&
!$OMP
iloop,iquit,ir,is,ithread,iwait,j,jd,jh,&
!$OMP           jr,k,k0,k00) &
!$OMP shared (fmatx,jmax,kmax,maxa,nthreads,ntotv,&
!$OMP         skmatx)

!         Factorize skmatx and reduce fmatx
      ithread=omp_get_thread_num()
      iloop=0
      iquit=0
      do while (iquit.eq.0)
        iloop=iloop+1
        j=(iloop-1)*nthreads+ithread+2
        if (j.gt.ntotv) then
          iquit=1
          exit
        endif
        ! Characteristic positions in skyline
        jr=maxa(j-1)
        jd=maxa(j)
        jh=jd-jr
        is=j-jh+2
        ! Start of core code
        ie0=0
```

```
!$OMP flush (kmax)
            dowhile (kmax.lt.jd)
            ! Initializations
              ihesitate=0
              ie0old=ie0
!$OMP flush (jmax,kmax)
            ! Judge if hesitation is necessary
            if (kmax.lt.jr) then
                ihesitate=1
                ie0=jmax
            endif
            if (jh.eq.2) then
                ! Reduce diagonal term
                iwait=1
            dowhile (iwait.eq.1)
!$OMP flush (kmax)
            if (kmax.ge.jr-1) then
                d=skmatx(jr+1)
                skmatx(jr+1)=d/skmatx(jr)
                skmatx(jd)=skmatx(jd)-d*skmatx(jr+1)
                iwait=0
            endif
            enddo
                ! Reduce right hand side (fmatx)
                fmatx(j)=fmatx(j)-skmatx(jr+1)*fmatx(j-1)
                ihesitate=0
            elseif (jh.gt.2) then
                ! Reduce all equations except diagonal
                ie=jd-1+(ie0-j+1)*ihesitate
                k00=jh-j-1+ie0old
                  if (k00.lt.0) k00=0
                    k0=0
                    do k=max0(jr+2,jd-j+ie0old+1),ie
                      ir=maxa(is+k0+k00-1)
                      id=maxa(is+k0+k00)
                      ih1=min0(id-ir-1,1+k0+k00)
                      if (ih1.gt.0) then
                        ih2=min0(id-ir-j+(j-1-k0-k00)&
                            *ihesitate,&
                            2-j+k0+k00+(j-1-k0-k00)&
                            *ihesitate)
                      if (ih2.lt.1) ih2=1
                      skmatx(k)=skmatx(k)&
                            -dot_product(&
                            skmatx(k-ih1:k-ih2),&

                      skmatx(id-ih1:id-ih2))
                      endif
                      k0=k0+1
                    enddo
                    if (ihesitate.eq.0) then
                       ! Reduce diagonal term
```

```
            ir=jr+1
            ie=jd-1
            k=j-jd
   do i=ir,ie
            id=maxa(k+i)
            d=skmatx(i)
            skmatx(i)=d/skmatx(id)
            skmatx(jd)=skmatx(jd)-d*skmatx(i)
   enddo

   ! Reduce right hand side (fmatx)
            fmatx(j)=fmatx(j)&
              -dot_product(skmatx(jr+1:jr+jh-1),&
                            fmatx(is-1:is+jh-3))
   endif
   endif
   if (ihesitate.eq.0) then
   !$OMP critical
   if (j.gt.jmax) then
            jmax=j
            kmax=jd
   endif
   !$OMP end critical
   endif
   !$OMP flush (jmax,kmax)
   enddo
   enddo
   !$OMP end parallel

   !        Divide by diagonal pivots
   do i=1,ntotv
            id=maxa(i)
            fmatx(i)=fmatx(i)/skmatx(id)
   enddo
   !        Back substitution
            j=ntotv
            jd=maxa(j)
100         d=fmatx(j)
            j=j-1
```

```
if (j.le.0) return
    jr=maxa(j)
if (jd-jr.gt.1) then
    is=j-jd+jr+2
    k=jr-is+1
    fmatx(is:j)=fmatx(is:j)-skmatx(is+k:j+k)*d
endif
    jd=jr
goto 100

return

  endsubroutine skyline_gauss_omp
```

Index

A
Anisotropy, 27

C
Computing
  domain decomposition, 67
  iterative solver, 67, 70
  parallelization, 67, 69, 113
  skyline solver, 68, 70, 72, 113
Contact
  electrical, 46
  frictional, 39, 45
  frictionless, 43
  resistance. *See* Material characterization
  sticking, 45
  thermal, 46
  tool, 38
Coupling procedures, 19

D
Domain integration, 24
Dynamic formulation, 5, 7

E
Elasticity, 32
Electrical model. *See* Electricity
Electricity, 18, 31
Elements
  contact, 41
  hexahedral, 3, 20, 38, 51
  tetrahedral, 51
  triangular, 38, 52

F
Flow formulation, 1, 5, 11
Friction characterization, 81
Friction laws, 37, 81

G
Gauss integration. *See* Domain integration

H
Heat transfer, 16, 25

I
I-Form, 14, 38
Incompressibility, 15

J
Joule heating, 17, 18

L
Lagrange multipliers, 15, 40

M
Material characterization
  compression test, 79
  compression test at elevated
    temperatures, 83
  electrical contact resistance, 84
  friction, 81
  stack compression test, 81

C. V. Nielsen et al., *Modeling of Thermo-Electro-Mechanical Manufacturing*
Processes, SpringerBriefs in Applied Sciences and Technology,
DOI: 10.1007/978-1-4471-4643-8, © The Author(s) 2013